AF812235

NO GRID ULTIMATE SURVIVAL BIBLE

Produce Everything You Will Ever Need on Your Property

2025 Module

Michael Mils | Daniel Grant PhD

| Patrick Gallagher | Smith Dyy

@HarbourHouse Press
harbourhousepublishing.com

Harbour
HOUSE PUBLISHING LTD

Acknowledgments

To the generations who lived by their hands, their wits, and their courage — thank you for the legacy.

To my family and supporters — thank you for believing in this vision.

And to every reader ready to reclaim their independence: This book was written for you.

Preface

You don't have to be an expert to take control of your future.
You only need the willingness to start — and the right guide.

Everything You Need On Your Property shows you how to transform your land, your skills, and your mindset into a complete off-grid system for food, water, energy, shelter, and security.

Every project in this book is designed to be practical, affordable, and achievable — no matter where you live or how much experience you have.

The tools for self-reliance are at your fingertips. It's time to build your freedom.

Disclaimer

Every piece of information provided in this book is officially intended for educational purposes only. It is important to note that a lot of efforts and research has been put in place to ensure they are safe, useful and impactful to the readers. However, the author is not to be held accountable for every form of loss, damage, injury etc resulting from the misuse or misapplication of the information in this publication.

Hence, readers are advised to do their due diligence and seek professional consultations especially in areas regarding construction, firearm, energy systems and herbal medicine.

Always prioritize safety, sustainability, and personal responsibility.

Introduction:

Building A Self-Sufficient Life

Imagine waking up every morning knowing that everything you need — water, food, energy, shelter, security — is right there on your own property. No dependency on fragile supply chains. No reliance on distant utilities or unpredictable services. Just pure, empowered self-reliance.

Everything You Need On Your Property is your blueprint for achieving true independence. In a world of increasing uncertainty, this book offers a clear, practical path to creating a self-sufficient oasis. It goes beyond survival; it is about thriving by building systems that work together seamlessly to support your daily needs.

The projects in this guide are not isolated tasks. Each one forms part of a larger, integrated strategy where water systems feed gardens, renewable energy powers workshops, greenhouses extend food production, and waste products become new resources. Through careful design, you can create a property that is not only functional but regenerative — a living, breathing ecosystem that sustains you and your family.

Following the philosophy found in *NO GRID Survival Projects: How to Produce Everything You Need on Your Property*, this book emphasizes:

- ✓ **Synergy Between Projects**: We'll show you how the byproducts of one project can fuel another. Rainwater captured from your roof can irrigate your survival garden. Solar energy can power your off- grid washing machine. Nothing goes to waste; everything has a purpose.
- ✓ **Sustainable, Local Material Use**: Each guide highlights ways to source building materials locally, reducing costs and increasing your independence. You'll learn to favor materials that are renewable, durable, and environmentally friendly.

- ✓ **Step-by-Step Practicality**: Each project provides detailed, easy-to-follow instructions, along with explanations of *why* it benefits your overall self- sufficiency. You'll never wonder if a project is worth the effort — you'll understand its role in the greater system.

✓ **Integration Tips**: Wherever possible, you'll find advice on connecting projects together. Think beyond "building a rain barrel" — think "building a water independence system."

Everything You Need On Your Property empowers you to take control of your life, protect your loved ones, and live with the freedom that comes from knowing your property provides for all your basic needs.

The skills you build here are timeless. The resilience you gain is priceless. Let's get started — your journey to true independence begins today.

Table of Content

I. Water, Energy & Environmental Solutions

A. Water Conservation & Collection

Water is life — especially in an off-grid environment where every drop must be harvested, stored, and used wisely. Effective water conservation and collection strategies form the cornerstone of any self-reliant property.

Rather than seeing rain, runoff, and household graywater as waste, off-grid living demands that we treat all water sources as valuable assets. Capturing rainfall, recycling used water, and reducing unnecessary consumption transform a property from vulnerable to resilient.

The projects in this section focus on:

- **Reducing Waste**: Simple systems like DIY sinks and gray-water collection capture and reuse water that would otherwise be lost.

- **Harvesting Nature's Gifts**: Rainwater catchment systems gather free, clean water that can sustain gardens, animals, and households.
- **Building Storage Capacity**: Proper tanks and reservoirs allow you to hold enough water to outlast dry spells and supply essential needs.

By linking these systems together — sink to graywater collection to garden irrigation — you create a fluid, efficient network that turns your property into a water- secure oasis.

With thoughtful design, sustainable materials, and strategic placement, you can transform water scarcity into water abundance — laying the groundwork for a thriving, independent homestead.

Chapter 1:
DIY Sink to Save Water

Water is one of your most precious resources when living off-grid. A DIY sink system allows you to use water efficiently, collect gray-water for garden use, and reduce overall waste. It's a simple project with major benefits for self-reliance.

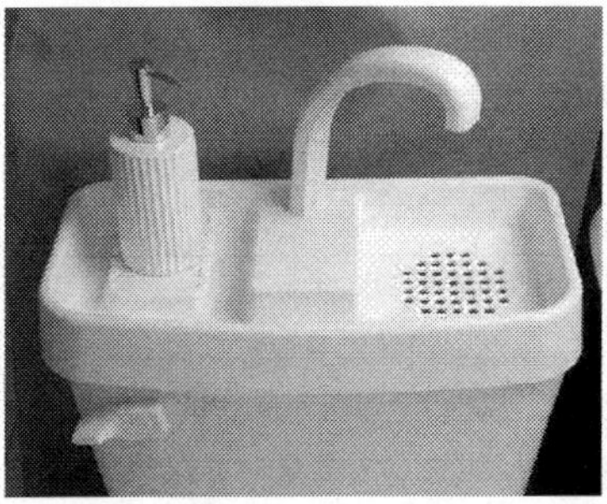

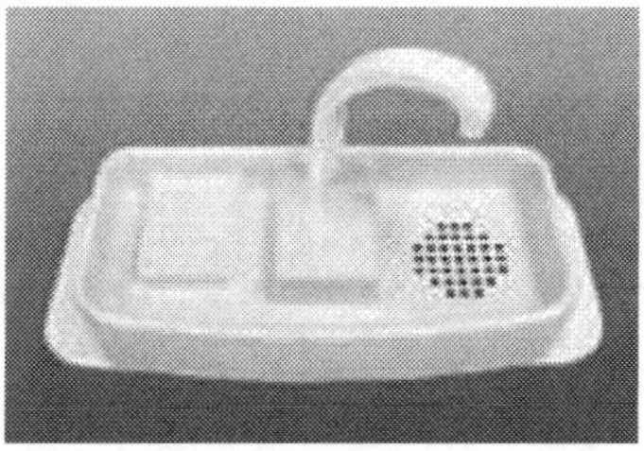

Materials Needed:

- A large plastic or metal basin
- PVC pipes (1-inch diameter)
- Faucet kit (manual or foot-pump type)
- 5-gallon collection bucket or drum
- Drain hose
- Hose clamps
- Sealant (silicone caulk)
- Screws, brackets, and basic tools (drill,

wrench, etc.)

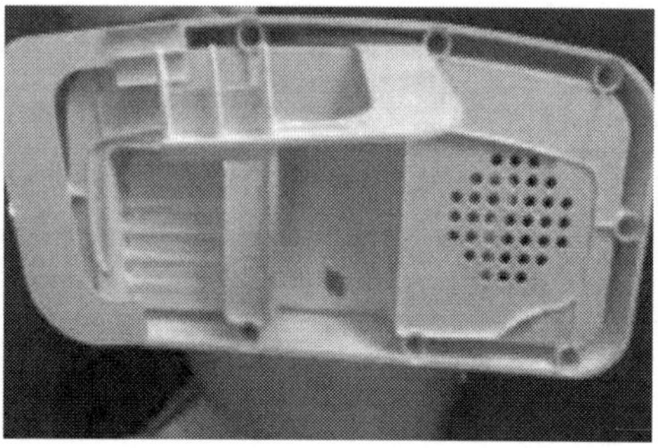

Instructions:

1. **Position Your Basin:**

   - Choose a sturdy table or build a simple wooden frame to mount your basin.

   - Ensure the location is near your garden, outdoor kitchen, or living quarters.

2. **Install the Faucet:**

   - Drill a hole near the back of the basin for the faucet installation.
   - Insert and secure the faucet following manufacturer instructions.

- If using a foot-pump faucet, mount the foot pedal at an accessible location.

3. **Create the Drain System:**

   - Drill a second hole at the bottom or side of the basin for drainage.
   - Attach a short piece of PVC piping to direct water into the collection bucket.
   - Secure connections with hose clamps and silicone sealant to prevent leaks.

4. **Set Up the Collection Bucket:**

   - Place the bucket beneath the drain pipe to capture used gray-water.

   - Install a lid with a hole cut for the drainpipe if you wish to reduce evaporation and contamination.

5. **Testing and Adjustments:**

   - Fill the basin and test the faucet and drainage system.

- Tighten any loose fittings and apply additional sealant if needed.

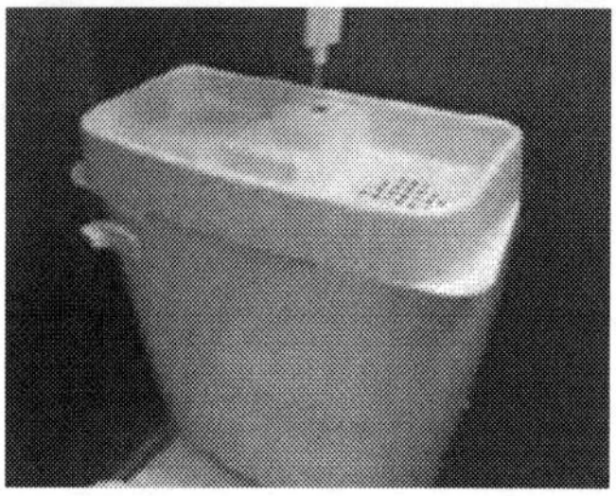

Integration Tip: The collected graywater can be filtered lightly and used to irrigate your survival garden or orchard. In an off-grid system, every drop counts! Pair this project with your rainwater catchment system to create a complete water independence network.

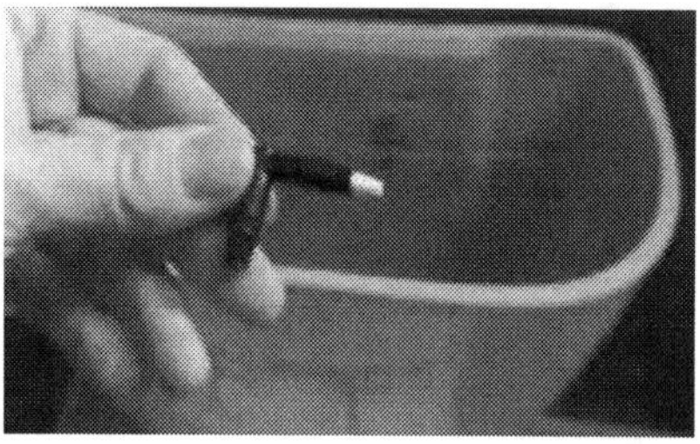

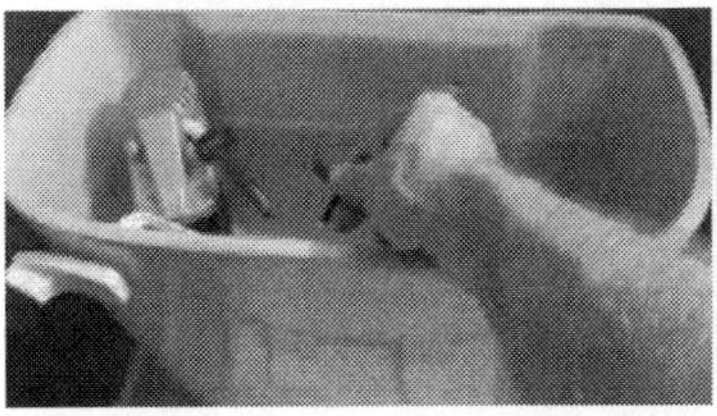

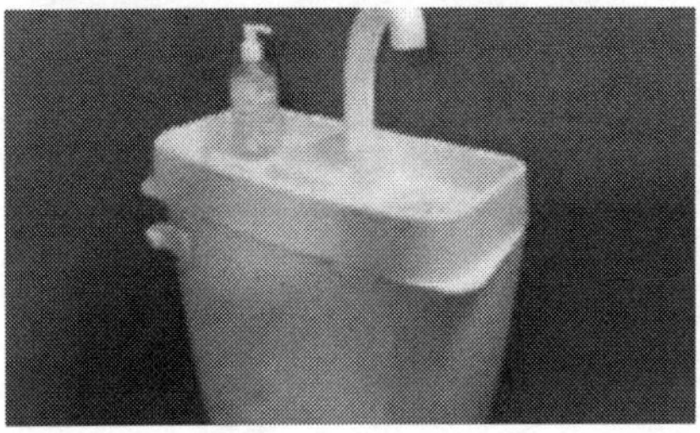
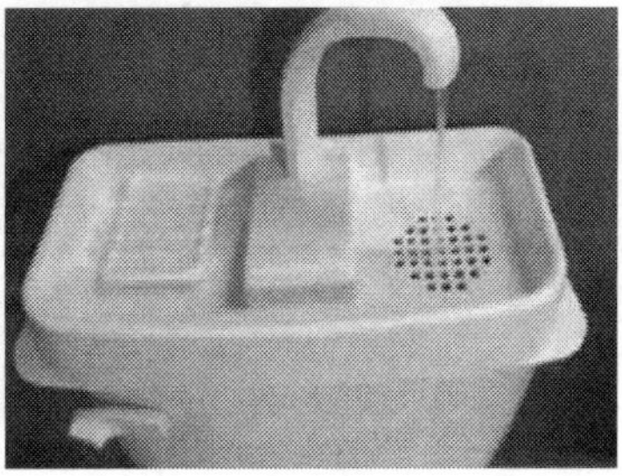

Benefits:

- Reduces water waste
- Supplies additional irrigation water
- Teaches responsible water management
- Low-cost build using mostly repurposed or easily sourced materials

Sustainability Focus: Try sourcing an old sink or washbasin from second-hand stores, construction surplus yards, or community "free-cycle" groups. PVC piping can often be found as leftover scraps from plumbing projects.

Chapter 2:
How to Build a Rainwater Catchment System

Overview: A rainwater catchment system allows you to capture free, abundant water directly from the skies.

Properly designed, it will dramatically increase your property's water independence and help sustain your gardens, animals, and household needs year-round.

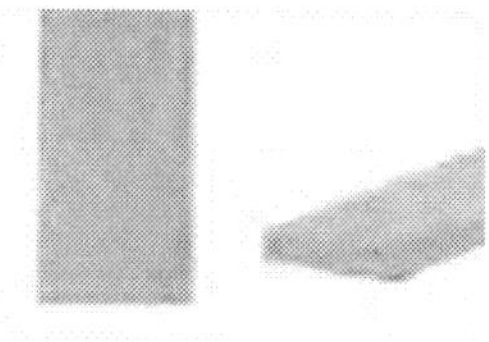

Materials Needed:

- Gutters and downspouts
- First flush diverter (optional but recommended)
- Fine mesh screen (for debris and mosquito control)

- 55-gallon barrels or larger tanks
- PVC piping and fittings (1- to 2-inch diameter)
- Hose bib/spigot kit
- Cinder blocks or sturdy stand
- Sealant (silicone caulk)
- Basic tools (saw, drill, level)

Building Process:

1. **Gutter and Downspout Setup:**

    - Install gutters along the edge of

your roof if they aren't already in place.

- Direct downspouts toward your designated rain barrel or tank location.

2. **Install First Flush Diverter (Optional):**

- A first flush diverter captures the initial runoff that may contain roof debris or contaminants. Install it on your downspout before it enters the catchment tank.

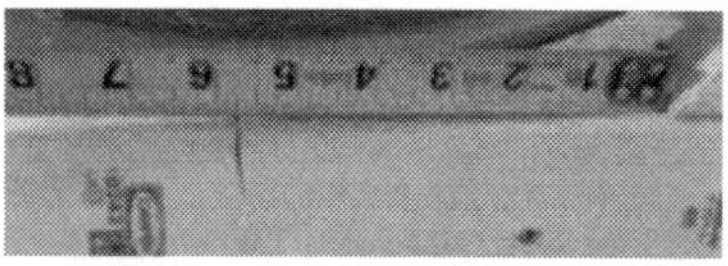
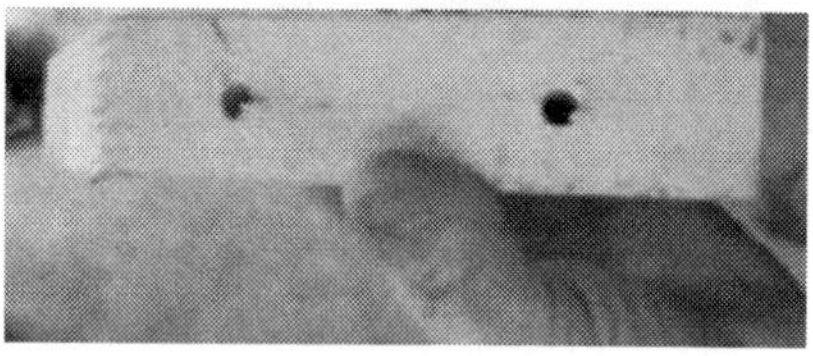
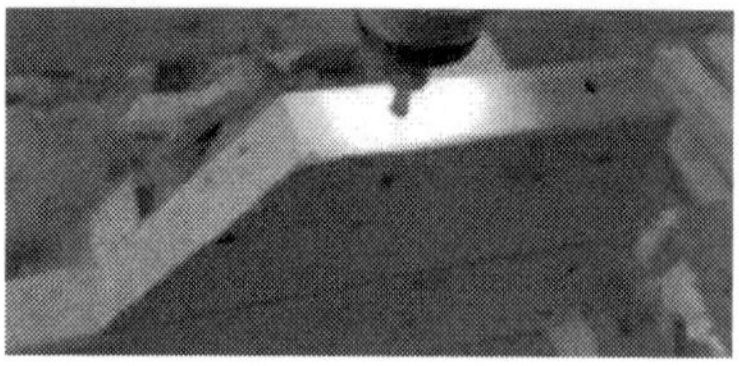

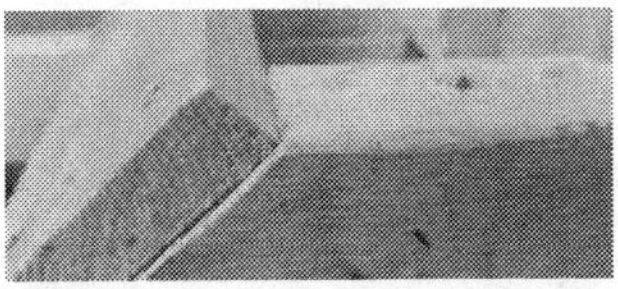

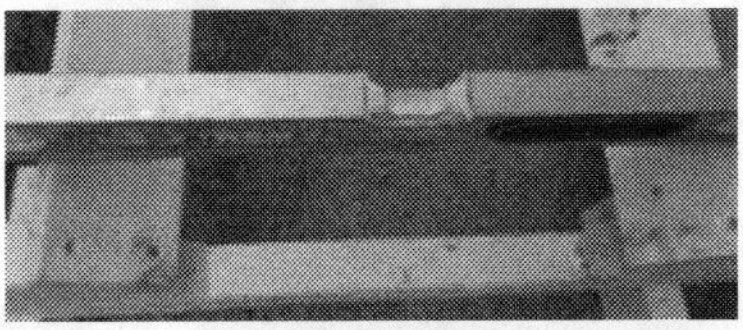

3. Prepare the Storage Container:

- Drill an inlet hole at the top of your barrel or tank to connect the downspout.
- Drill a hole near the bottom to install the hose bib or spigot for water access.
- Attach a fine mesh screen over the inlet hole to block debris and mosquitoes.

4. **Position the Container:**

- Place the container on a raised stand (such as cinder blocks) to improve water flow from the spigot.
- Ensure the platform is level and stable to support the full weight of the container.

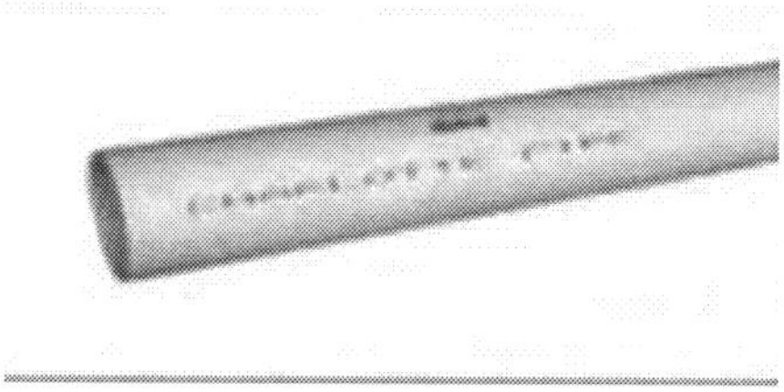

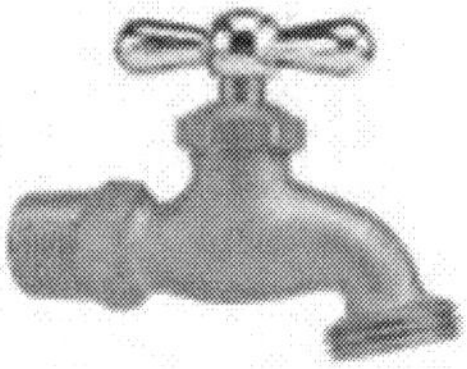

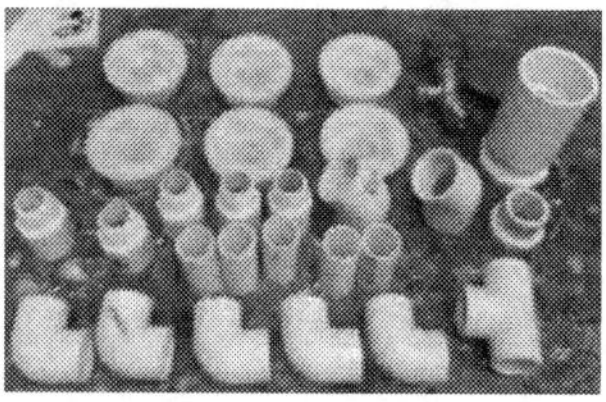

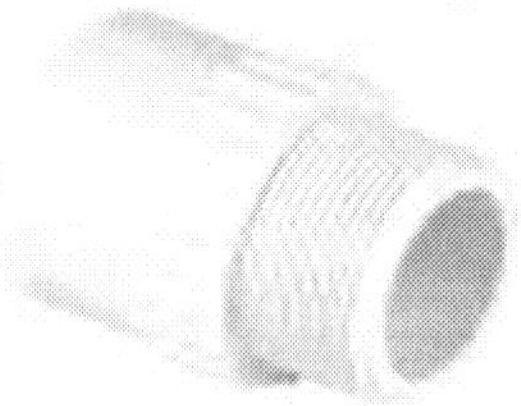

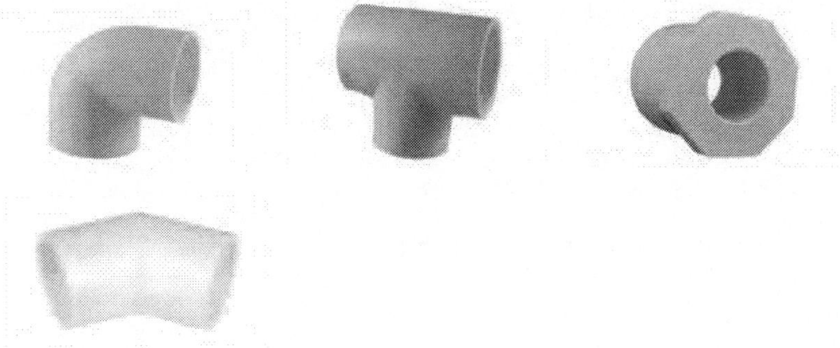

5. **Connect the Downspout:**

   - Secure the downspout into the inlet hole of the barrel, using sealant to prevent leaks.

6. **Overflow Management:**

   - Drill an overflow hole near the top of the container and attach a hose or pipe to

     direct excess water safely away from your structure.

Placement of the Rainwater Catchment:

- Locate your system near high water-need areas like gardens or animal pens.
- Consider shading the container to reduce algae growth.
- If possible, daisy-chain multiple barrels to increase storage capacity.

Integration Tip: Use your rainwater catchment to feed your DIY sink, irrigation systems, and even your solar- heated water projects. Water that falls freely from the sky should never go to waste — when captured and stored correctly, it becomes one of your strongest assets for full off-grid living.

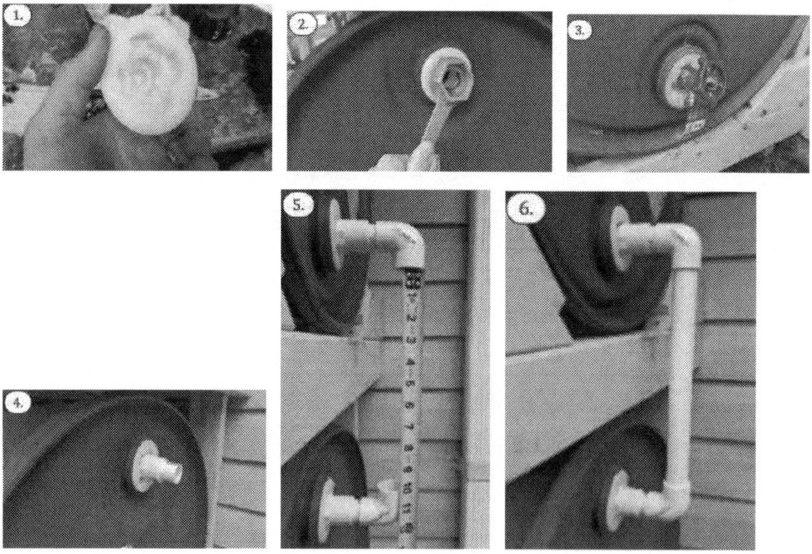

Benefits:

- Provides free, renewable water
- Reduces dependency on wells, city water, or delivered supplies
- Supports gardens, animals, and household tasks
- Simple, low-cost design scalable to any property size

Sustainability Focus: Repurpose food-grade barrels whenever possible, and prioritize reusing gutters and downspouts salvaged from old buildings or remodeling projects.

Chapter 3:
How to Build a Water Tank for Long-Term Storage

Short-term water storage is essential, but true off-grid resilience demands the ability to store large quantities safely over long periods. A DIY water tank project ensures a reliable reserve for emergencies, droughts, or seasonal shortages.

Determining Capacity: "How Much Water?"

- A general rule is to store at least one gallon per person per day.
- Multiply your daily needs by 30, 60, or 90 days depending on your preparedness goals.
- Factor in water needs for gardens and livestock if applicable.

Container Options:

- **55-Gallon Drums**: Affordable, modular, and easy to transport.
- **Jerry Cans**: Great for portability but less efficient for large-scale storage.

- **Soda Bottles**: Good for small backup reserves, but not ideal for primary storage.
- **Disposable Bottles**: Only for temporary use; prone to degradation over time.

Design Considerations:

- Use food-grade, UV-resistant containers whenever possible.
- Store in a shaded, cool location to extend water quality.
- Elevate tanks on platforms to allow easy

gravity-fed dispensing.

Sanitizing Procedures:

1. Rinse the container with clean water.
2. Fill with a solution of one teaspoon of unscented bleach per gallon of water.
3. Shake and roll the container to coat all surfaces.
4. Let sit for 30 minutes, then rinse thoroughly.

Building Steps:

1. **Select and Sanitize Containers:**
    - Clean all containers before use.
    - Double-check seals, lids, and spouts for integrity.

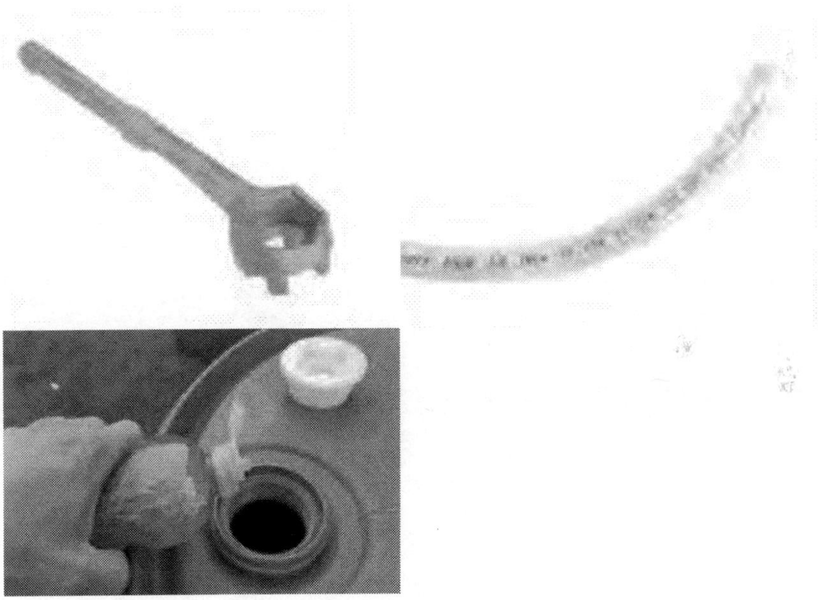

2. **Design Platform:**
    - Build a sturdy wooden or cinder block stand.
    - Ensure it can handle the full weight of a filled tank (about 8.34 lbs per

gallon).

3. **Assembly:**

    - Stack or cluster containers securely.
    - Use PVC plumbing to connect tanks if creating a multi-tank system.
    - Install spigots or valves for easy access.

Integration Tip: Connect your rainwater catchment system directly to your long-term storage tanks. During rainy

seasons, you can fill your reserves without lifting a finger, creating a closed-loop water system.

Benefits:

- Provides critical backup in emergencies
- Reduces vulnerability to seasonal changes
- Ensures water availability for gardens, animals, and household use

Sustainability Focus: Whenever possible, reclaim containers from food processing industries or agricultural suppliers. A strong off-grid property uses "waste" from other industries to create lasting infrastructure.

B. Water Treatment & Filtration

Collecting and storing water is only the first step in achieving true water independence. Ensuring that your water is safe to drink, cook with, and bathe in is just as critical. In an off-grid environment, the responsibility for water purity falls entirely on you.

Water treatment and filtration projects transform your stored rainwater, well water, or even harvested graywater into a safe, usable resource. These systems protect you from bacteria, viruses, heavy metals, and debris — ensuring health and resilience across your entire homestead.

Just like your storage and catchment systems, water purification should be layered and integrated:

- **Primary Filters** remove sediment and large particulates.
- **Secondary Treatments** like charcoal filtration eliminate contaminants and improve taste.
- **Emergency Methods** (boiling, chemical treatments, solar disinfection) act as critical

backups.

We will explore DIY filtration systems, basic hand pumps, and even methods to pull moisture directly from the air.
Every project is designed not just to solve a problem but to strengthen your entire off-grid water independence network.

With smart design and simple materials, you can ensure that every drop you collect becomes a resource you can trust — even when utilities fail or emergencies strike.

Chapter 4:
Designing a Water Storage System

A water storage system is the backbone of any off-grid property. While individual tanks and barrels provide immediate storage, a fully thought-out system ensures reliable, scalable, and sustainable water availability year- round. By designing a system with redundancy, accessibility, and integration in mind, you create a truly resilient homestead.

Preparing & Sanitizing Containers: Before integrating any container into your system, proper cleaning and sanitation is critical to prevent algae growth, bacterial contamination, and unpleasant odors.

Sanitizing Steps:

1. Rinse each container thoroughly with clean water.
2. Prepare a sanitizing solution: Mix 1 teaspoon of unscented household bleach per gallon of water.
3. Fill the container about 1/4 full, seal, and

vigorously shake to cover all surfaces.

4. Let the solution sit in the container for 30 minutes.
5. Empty and rinse thoroughly with potable water.

Choosing Between Vertical and Horizontal

Placement: Vertical Tanks:

- **Pros:** Smaller footprint, easier gravity-fed pressure, better for limited space.
- **Cons:** May require more sturdy foundations due to concentrated weight.

Horizontal Tanks:

- **Pros:** Greater stability on uneven ground, easier to link multiple tanks.
- **Cons:** Require more surface area, may need more complex piping for pressure.

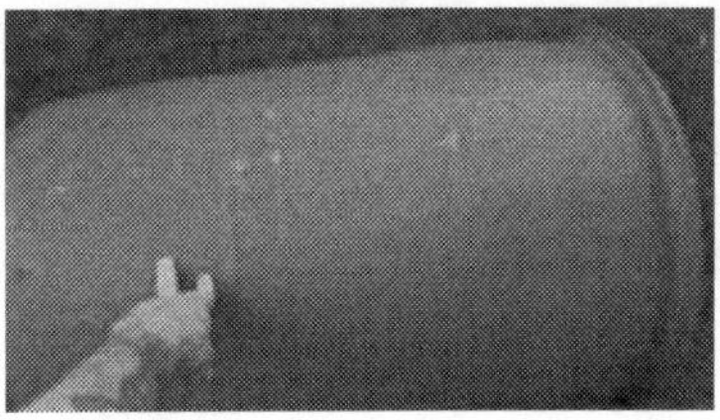

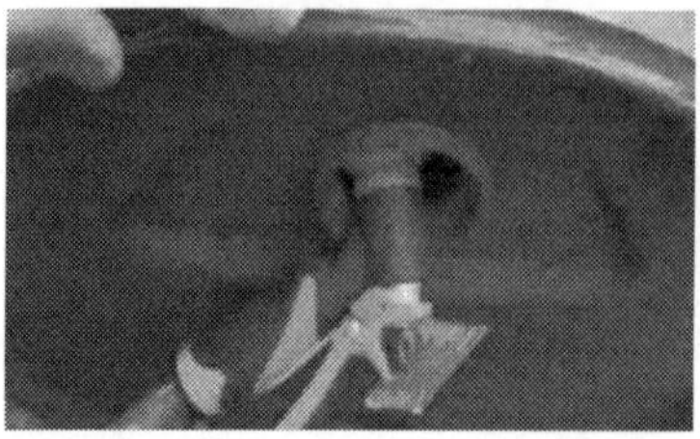

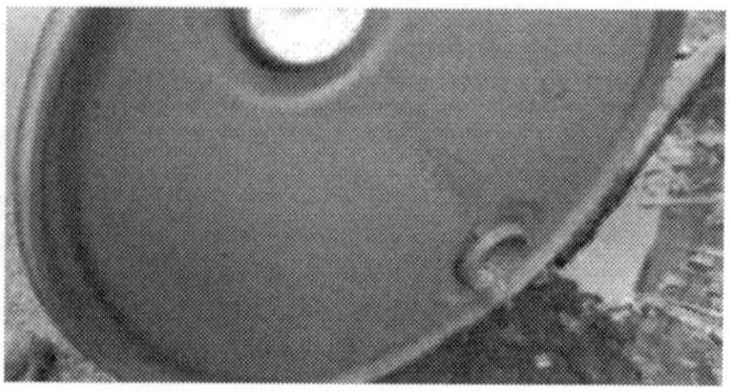

Design Considerations:

1. **Gravity Flow vs. Pump-Assisted:**
    - Elevate tanks on platforms to use gravity for water pressure.
    - Pumps offer pressure boosts for irrigation or household plumbing but require energy input.

2. **Accessibility:**
    - Install valves or spigots at a comfortable working height.
    - Include overflow outlets and easy drain points for cleaning.

3. **Modularity:**
    - Design your system so you can add tanks as your needs grow.
    - Use quick-connect PVC fittings

or flexible hose systems for future expansion.

4. **Redundancy:**
    - Build in backups. Two smaller tanks are often better than one giant one in case of leaks or failures.

5. **Filtration and Pre-Treatment:**
    - Install a basic sediment filter before storage.
    - First-flush diverters and mesh screens keep contaminants out.

Building Steps:

1. **Plan Your Layout:**
    - Sketch your property and decide the best locations based on catchment sources, elevation, and proximity to gardens and buildings.

2. **Foundation Building:**
    - Use compacted gravel pads, concrete blocks, or wooden

stands to stabilize tanks.
- Ensure level placement to avoid tipping and stress fractures.

3. **Plumbing Installation:**
    - Connect tanks with PVC piping and ball valves.
    - Install backflow preventers to keep water from contaminating the source.

4. **Overflow and Emergency Drains:**
    - Position overflow hoses away from structures.
    - Emergency drains allow for flushing the system during maintenance.

Integration Tip: Design your water storage to work hand- in-hand with your rainwater catchment, graywater collection, and even solar water heating projects. The more systems you link together, the more your property becomes a living, water-resilient machine.

Benefits:

- Ensures a steady supply of clean water
- Allows for scaling up over time
- Reduces vulnerability to seasonal shortages or disasters
- Supports irrigation, livestock, cleaning, and emergency preparedness

Sustainability Focus: Use recycled, food-grade containers and reclaimed plumbing materials wherever possible. Look for tanks from agricultural surplus sales, food processing

plants, or local salvage yards to keep costs low and environmental impact minimal.

Chapter 5:
DIY Water Filter Construction

No matter how clean your water looks, unseen contaminants can lurk within — bacteria, viruses, chemical residues, or microscopic debris. Off-grid living demands that you take full responsibility for the safety of your water supply.

A homemade water filter provides a vital layer of protection, transforming rainwater, well water, or stored supplies into safe, drinkable water. This project is affordable, highly effective, and adaptable to various supply situations, giving you peace of mind no matter the source.

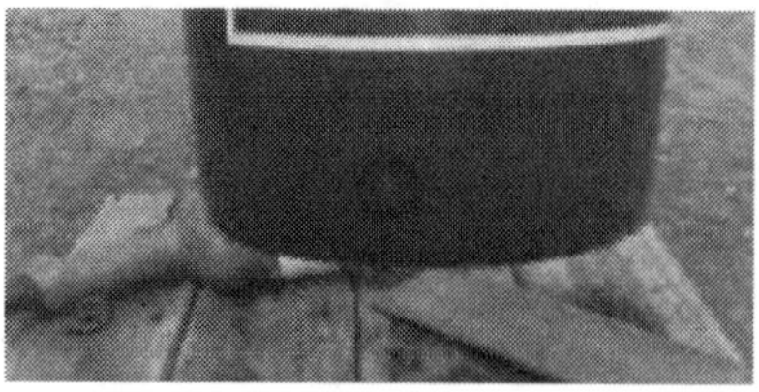

With simple materials like buckets, sand, gravel, and activated charcoal, you can build a filter capable of removing major impurities and improving water taste and quality.

Proper filtration is the bridge between survival and thriving. When combined with catchment systems, storage tanks, and purification methods like boiling or solar

disinfection, a DIY filter becomes an essential part of your water resilience network.

Materials Needed:

- Two food-grade 5-gallon buckets with lids
- Spigot kit or bulkhead fitting (for bottom bucket)
- Plastic tubing (optional, for overflow or draining)
- Fine-grit sand (play sand or masonry sand)
- Coarse gravel
- Activated charcoal (aquarium-grade or homemade)
- Cotton cloth or coffee filters
- Drill and hole saw attachment
- Knife or scissors
- Silicone sealant

Step-by-Step Assembly:

1. **Prepare the Buckets:**
    - Drill a hole near the bottom of one bucket for the spigot.

- Install the spigot according to manufacturer instructions, sealing around it with silicone.

2. **Create the Filter Layers:**
    - Cut a piece of cotton cloth to fit the bottom of the top bucket.
    - Layer as follows (bottom to top):
        1. Cotton cloth
        2. 2-3 inches of activated charcoal
        3. 2-3 inches of fine sand
        4. 2-3 inches of coarse gravel
    - Repeat layering if bucket depth allows for double layers, increasing filtration efficiency.
3. **Assemble the System:**
    - Drill a small hole in the lid of the top bucket to allow water pouring without removing the lid (optional).
    - Stack the top (filter) bucket onto the bottom (collection) bucket securely.
4. **Filtering Water:**
    - Pour raw water into the top bucket slowly.
    - Allow water to percolate through the layers and collect in the bottom bucket.
    - Use the spigot to dispense filtered water.

Care and Maintenance:

- Rinse the filter layers lightly if they become clogged with debris.
- Replace the activated charcoal every 3-6 months or after heavy usage.
- Sanitize buckets and spigot periodically with a diluted bleach solution.

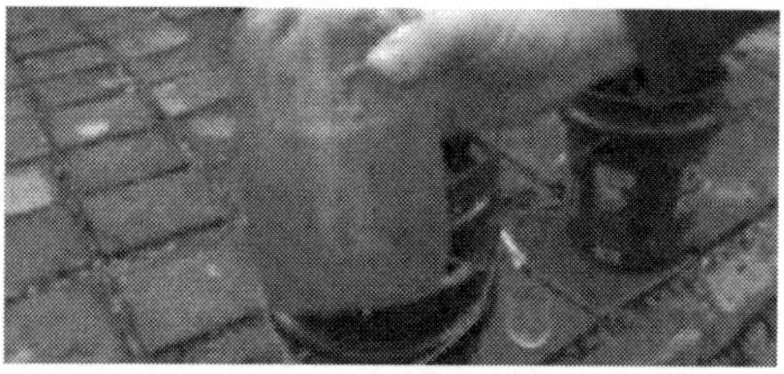

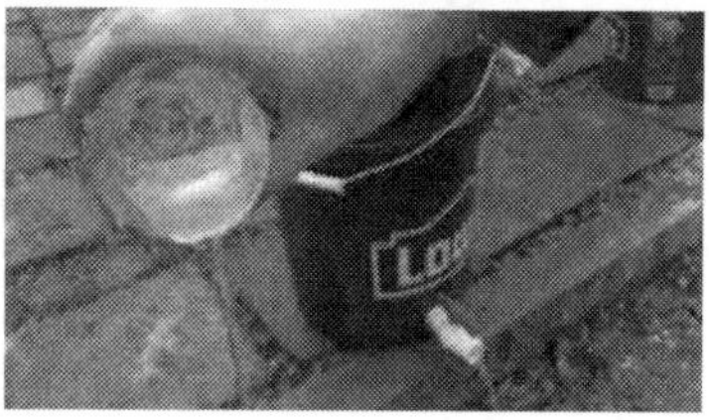

Integration Tip: Position your DIY water filter next to your rainwater catchment tanks or long-term storage barrels. This allows you to immediately process collected water into potable supplies without moving heavy containers around your property.

Benefits:

- Removes sediment, debris, and many harmful contaminants
- Enhances water taste and odor
- Extremely low-cost compared to commercial filters
- Simple repairs and maintenance using basic tools and materials

Sustainability Focus: Wherever possible, use salvaged or repurposed food-grade buckets, and consider making your

own activated charcoal using hardwood scraps and controlled burning techniques.

Chapter 6:
How to Build a Hand Pump

When power grids fail or automated systems break down, having a reliable manual method to access water can make the difference between hardship and security. A DIY hand pump allows you to draw water from wells, tanks, or cisterns without relying on electricity.

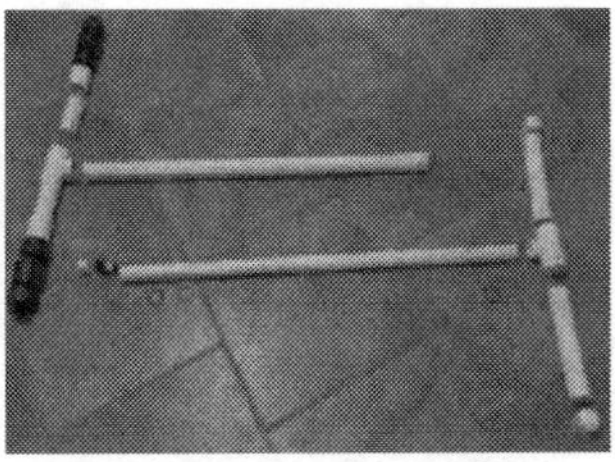

Simple yet powerful, hand pumps are a cornerstone of resilient off-grid living. With a few basic materials and some ingenuity, you can create a system that taps into underground or stored water sources anytime you need it.

This project emphasizes low-cost, locally sourced materials and teaches basic mechanics that empower you to build, repair, and adapt your pump for decades to come.

Materials Needed:

- 1" to 1.5" diameter PVC pipe (length depends on water depth)
- 0.5" PVC pipe (for plunger rod)
- PVC foot valve or one-way check valve
- Rubber or leather washer (for plunger)
- PVC end caps
- T-handle or standard hand grip (wood or PVC)
- Hose clamps
- Primer and PVC cement
- Drill, saw, wrenches

Plunger & Valve Assemblies:

1. **Build the Plunger:**
   - Cut a circular rubber or leather washer to fit snugly inside the main PVC pipe.
   - Attach the washer to a short section of 0.5" PVC rod using a nut and bolt.
   - This assembly acts as a piston, forcing water upwards as you pump.
2. **Install the Foot Valve:**
   - Attach a store-bought or DIY one-way valve to the bottom of the main pipe.

- The foot valve allows water to enter but prevents it from draining back down.

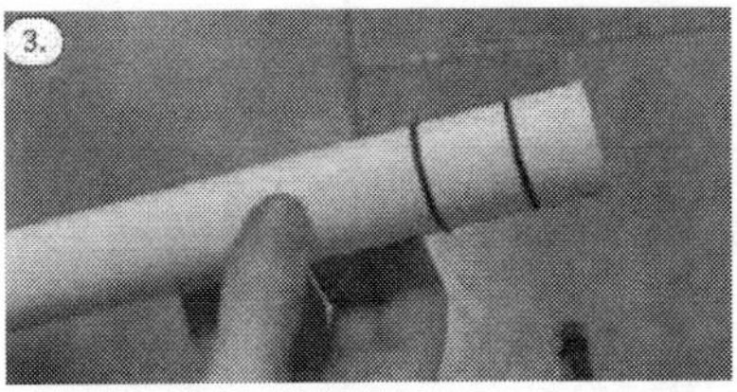

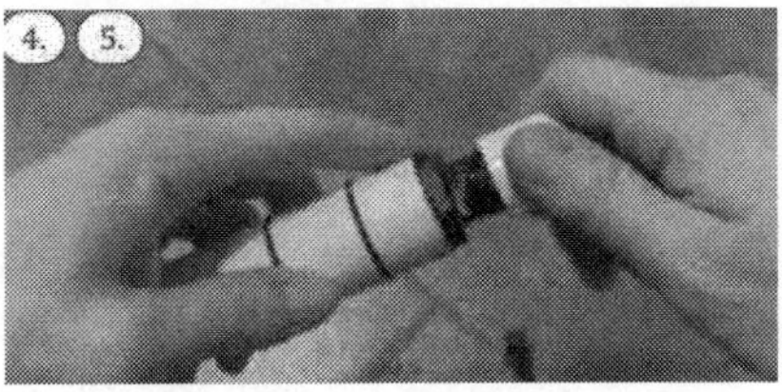

Building Process:

1. **Assemble the Pipe:**
   - Prime and cement the main length of 1-1.5" PVC pipe, ensuring it's long enough to reach your water source plus an extra 3-5 feet.

2. **Install the Plunger:**
   - Feed the plunger rod down through the main pipe.
   - Seal the top end with a PVC end cap, drilling a hole to allow the plunger rod to pass through while keeping pressure inside.
3. **Attach the Handle:**
   - Attach a simple T-handle or other comfortable grip to the top of the

plunger rod.

    - Secure firmly with screws or strong adhesive.

4. **Set Up the Pump:**
    - Insert the completed pump assembly into your water source.
    - Begin moving the handle up and down to draw water.

Tools, Materials, and Variations:

- **Shallow Wells (Less than 25 feet):** Basic hand pumps are highly effective.
- **Deep Wells:** May require specialized designs like deep well piston pumps or double-acting pumps.
- **Upgrade Options:** Use metal piping and professional-grade valves for longer service life if local materials permit.

Integration Tip: Connect your hand pump output to storage tanks or gravity-fed distribution systems. This enables you to manually refill water reserves during outages or emergencies, ensuring continuous supply for gardens, livestock, and household needs.

Benefits:

- Provides water access without electricity
- Extremely low cost compared to commercial pumps
- Simple construction and easy repair
- Portable and adaptable to different water sources

Sustainability Focus: Whenever possible, use reclaimed PVC pipe and fittings from construction leftovers.

Repurpose old bicycle handles or wooden rods for pump handles to minimize waste and lower costs.

Chapter 7: Making Drinkable Water from Air

Water is all around us, even when we can't see it. Every breath you take contains moisture, and with the right tools, you can harvest this hidden resource — a critical skill during droughts, grid failures, or long-term emergencies.

Atmospheric water generation sounds high-tech, but at its core, it uses simple principles: cooling warm air until the moisture condenses into liquid water. By understanding and adapting basic dehumidification, you can create small but steady water supplies off-grid.

This project explores how to use common appliances like dehumidifiers to extract water from the air safely, how to test the water for purity, and how to integrate this method into a complete survival water system.

Materials Needed:

- Portable dehumidifier (energy-efficient model preferred)
- Collection tank or clean container

- Activated carbon water filter (optional but recommended)
- Extension cord or solar generator (optional for off- grid power)
- Water testing kit (for pH, bacteria, and heavy metals)

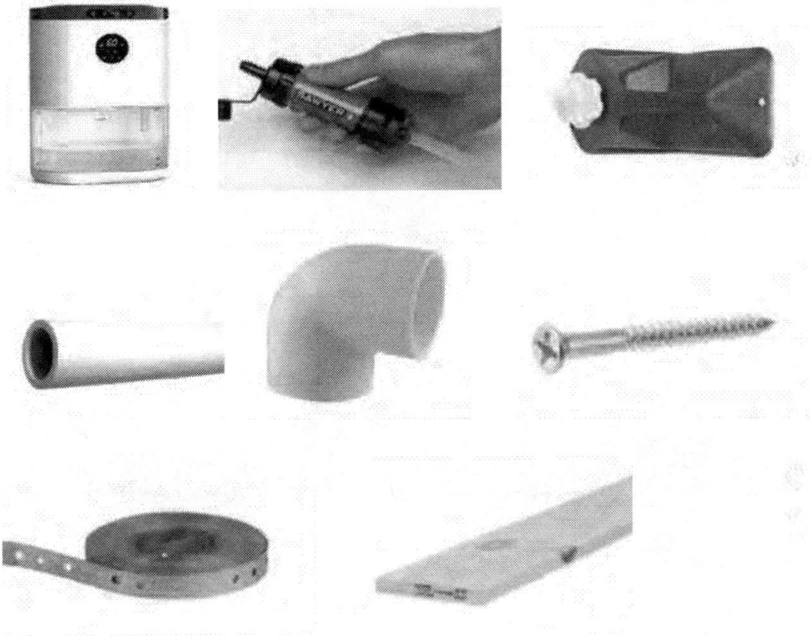

Safety of Dehumidifier Water:

Water collected from air may contain:

- Metal residues from the appliance coils
- Airborne contaminants like dust, mold spores, or bacteria

To ensure safe use:

- Always run collected water through a charcoal- based or ceramic filter.
- Perform regular water testing, especially if using water for drinking.
- Clean your dehumidifier coils and collection tank frequently with vinegar or hydrogen peroxide.

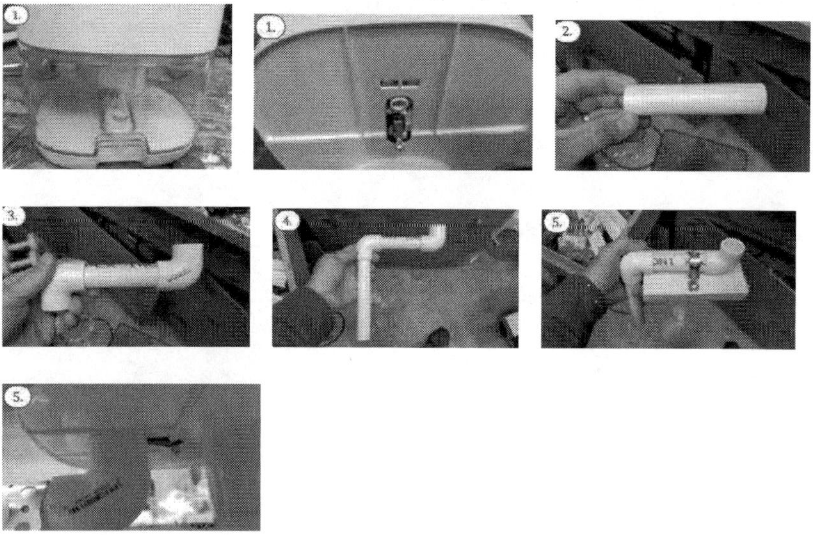

Filter Choices and Testing:

1. **Primary Filtration:**
   - Use an activated carbon filter to remove chemical residues, odors, and some biological contaminants.

2. **Secondary Safety Check:**
    - Use a simple home water testing kit to monitor for pH balance, bacterial presence, and heavy metals.
    - If unsafe, treat water by boiling or adding purification tablets.

Building Process:

1. **Set Up the Dehumidifier:**
    - Place the unit in a warm, humid location for maximum water extraction (greenhouses, kitchens, basements).
    - Ensure good airflow for efficiency.
2. **Direct Collection:**
    - Route the drain hose or drip line into a clean collection tank.
    - If no hose option is available, manually empty the dehumidifier reservoir into a tank daily.
3. **Filtration:**
    - Pass collected water through a simple DIY water filter (such as the charcoal bucket system described earlier) before drinking or using for sensitive tasks.
4. **Testing:**

- ○ Test water at least weekly if using regularly.
- ○ Adjust filtration or treatment methods based on test results.

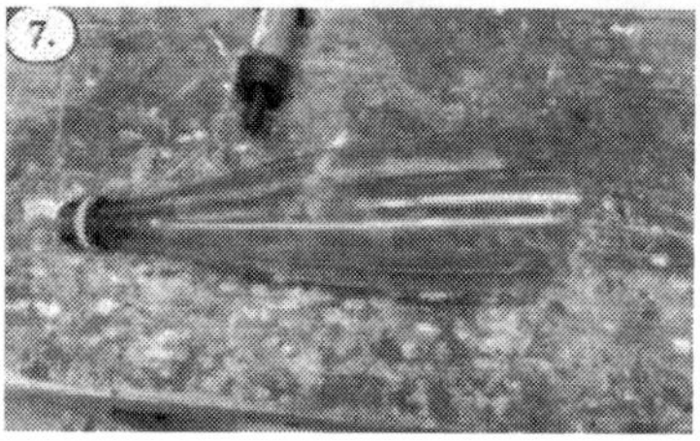

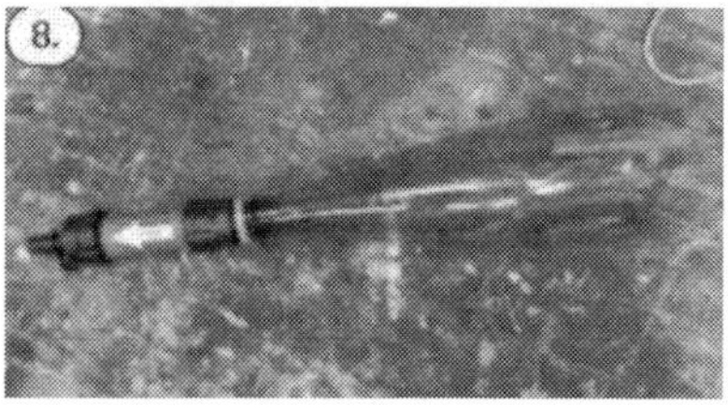

Integration Tip: Connect your dehumidifier output directly into your water storage system, especially during humid seasons when air moisture is high. If powered by solar panels or a bicycle generator, you create an independent "air well" that enhances your water security even during dry periods.

Benefits:

- Provides additional water source beyond rain or wells

- Reduces dependence on unpredictable rainfall patterns
- Great emergency backup method
- Can be powered off-grid with solar energy

Sustainability Focus: Whenever possible, choose Energy Star-rated dehumidifiers and power them with renewable sources. Even a small solar panel array can sustain a basic dehumidifier setup, making water-from-air harvesting fully off-grid and eco-friendly.

C. Renewable Energy & Power Generation

True off-grid living is not just about water and food security; it is equally about energy independence.
Electricity empowers everything from lighting and communication to refrigeration and water pumping. Without it, modern conveniences and survival systems quickly grind to a halt.

Renewable energy projects empower you to capture nature's gifts — the sun, the wind, flowing water, and even human muscle — to produce clean, sustainable power.

This section explores practical DIY methods to:

- Generate electricity from simple bicycle-powered systems
- Harness small streams for hydroelectric power
- Build affordable solar arrays
- Construct home-sized wind turbines

Each project is designed to be achievable with basic tools, scavenged materials, and a good dose of determination.

Integration is key: surplus solar power can charge

communication radios, run dehumidifiers, and store in batteries to operate pumps.

Building energy independence is not just about survival; it's about reclaiming your freedom, securing your family's future, and mastering the natural forces that surround you.

Let's dive in and bring power to your homestead — no utility bills required!

Chapter 8:
DIY Bicycle Generator

Your legs are a powerhouse! With just a bit of clever engineering, you can transform a regular bicycle into a reliable, renewable source of electricity. A bicycle generator taps into human mechanical energy and converts it into usable electric power — enough to charge batteries, run lights, or power small appliances.

This project is perfect for emergency backup power, fitness-energy integration, or just learning the basic principles of mechanical-to-electrical conversion. Best of all, it repurposes common items and builds your understanding of how energy truly works.

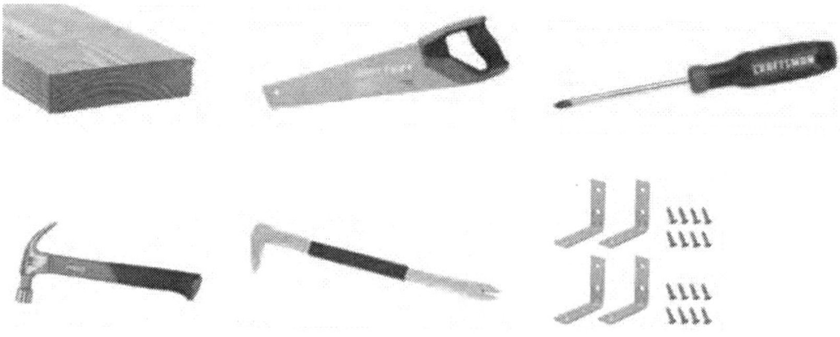

Materials Needed:

- Standard bicycle (mountain or road bike)
- Bicycle stand or DIY mounting frame
- DC motor (preferably 12-24V, brushed motor)
- Belt, chain, or friction roller system (depending on motor type)
- Rectifier (to convert AC to DC if needed)
- Charge controller (optional for battery systems)
- 12V deep-cycle battery or storage system (optional)
- Wires, connectors, and basic electrical tools
- Multimeter (for testing output)

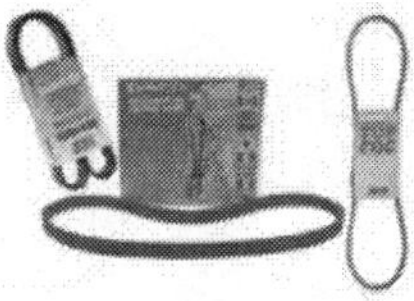

Building Process:

1. **Mount the Bicycle:**
   - Stabilize the bike on a stationary stand so the back wheel spins freely.
   - DIY Option: Build a frame from wood, metal, or PVC pipe to lift and secure the rear wheel.
2. **Connect the Motor:**
   - Position the DC motor so its shaft or friction roller presses firmly against the bicycle tire.
   - Tighten securely to maintain good contact during pedaling.
3. **Electrical Setup:**
   - Wire the motor output to the rectifier (if needed) to produce clean DC current.

   - Connect output to a battery or directly to a device through a voltage regulator.

4. **Testing:**
    - Spin the back wheel and measure voltage output with a multimeter.
    - Adjust motor tension or wheel contact if voltage is too low.
5. **Load Management:**
    - Use a charge controller to prevent overcharging batteries.
    - Install fuses or circuit breakers for safety.

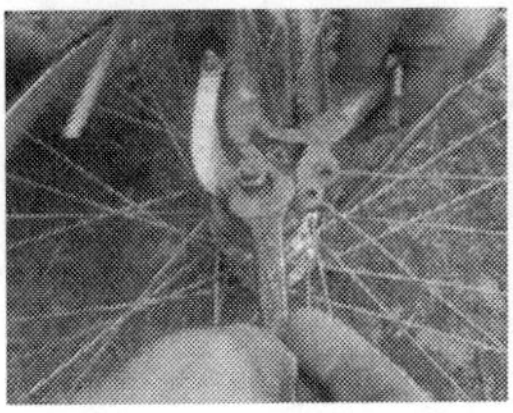

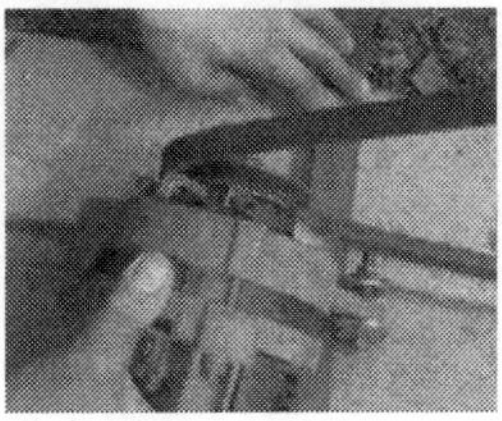

Variations and Upgrades:

- Use gear ratios to optimize pedaling ease and power output.
- Integrate USB ports for phone or tablet charging.
- Link multiple bikes together for larger systems (community gym models).

Integration Tip: Use your bicycle generator to power vital off-grid components: emergency radios, dehumidifiers, LED lighting systems, or water pumps. Human-powered energy acts as a perfect backup when solar, wind, or hydro sources are unavailable.

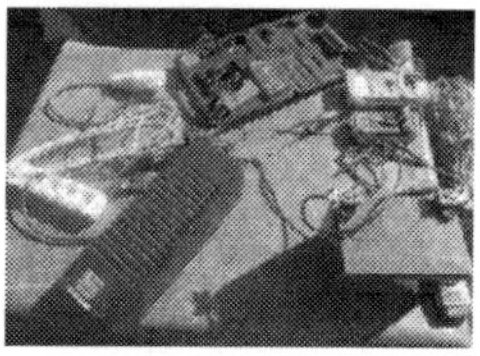

Benefits:

- Converts human effort into usable energy
- Excellent emergency and backup power source
- Easy to build using salvaged parts
- Great for fitness while generating electricity

Sustainability Focus: Scavenge old treadmills, washing machines, or scooters for motors. Build stands from scrap lumber or repurposed metal to minimize new material usage and keep the project low-cost and eco-friendly.

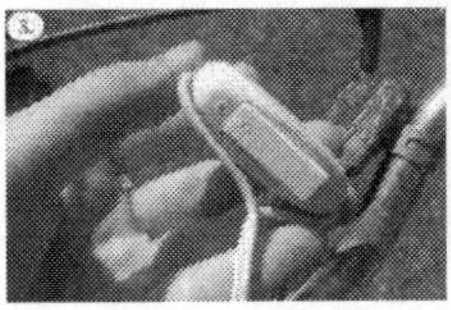

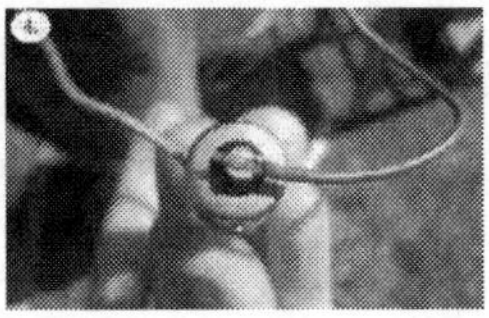

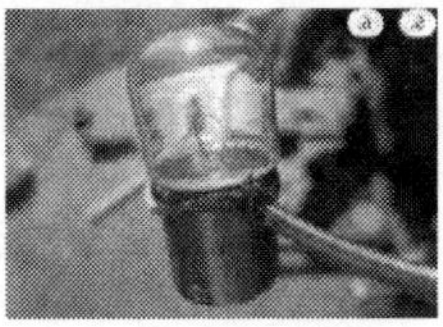

Chapter 9: How to Build a Hydroelectric Generator

Flowing water is one of nature's most powerful and reliable energy sources. Even a modest stream or creek on your property can be harnessed to generate continuous, renewable electricity with the right setup.

A small-scale hydroelectric generator offers huge advantages: it's low-maintenance, produces steady power day and night, and works regardless of weather conditions. Best of all, it can be built with affordable materials and basic skills.

This project will show you how to capture energy from moving water to power batteries, lights, communication systems, or essential appliances — strengthening your property's energy independence.

Materials Needed:

- Permanent magnet DC motor (preferred for easy energy capture)
- Waterwheel, paddle wheel, or turbine

(DIY or purchased)

- PVC piping (for penstock/ water delivery)
- Bearing housings and shaft
- Wooden frame or metal mounting stand
- Rectifier (if AC motor is used)
- Charge controller (for battery charging)
- 12V deep-cycle battery bank (optional)
- Electrical wires, connectors, and basic tools
- Multimeter

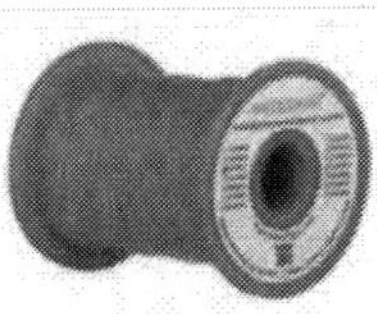

Tools Needed:

- Power drill
- Wrenches and screwdrivers
- Saw (for wood or metal frame construction)
- Level
- Shovels (for minor earthwork if necessary)

Building Process:

1. **Survey Your Site:**
    - Locate the spot with the strongest, fastest flow.
    - Measure "head" (the vertical drop) and flow rate; more head and more flow equals more power.

2. **Build the Water Delivery System (Penstock):**
    - Use PVC piping to direct water from a higher elevation toward your turbine or waterwheel.
    - Keep the piping as straight as possible to maximize pressure.
3. **Install the Waterwheel or Turbine:**
    - Mount the waterwheel securely on a shaft with bearings.
    - Ensure it spins freely and is properly aligned with the incoming water flow.
4. **Attach the Generator:**
    - Connect the shaft to your permanent magnet DC motor.
    - Use belts, chains, or direct coupling to transfer rotational energy.
5. **Electrical Setup:**

    - Wire the motor output to a rectifier (if needed) and then into a charge controller.
    - Connect to your battery bank or directly power DC devices.

6. **Testing and Fine-Tuning:**
    - Measure voltage and amperage with a multimeter.
    - Adjust waterwheel angle, flow direction, or belt tension to optimize output.

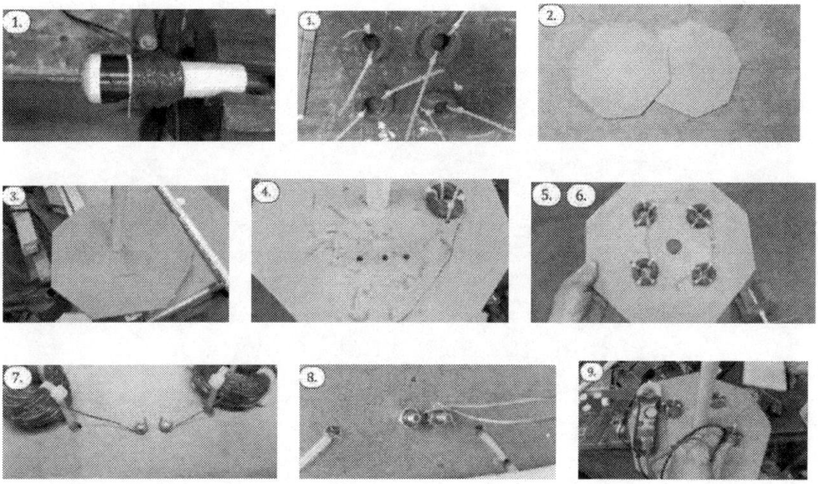

Considerations and Modifications:

- Protect the system from debris by installing a water intake screen.
- Anchor all components securely to withstand

floods or seasonal changes.

- Use waterproof electrical connections and housing to prolong component life.

Safety Measures:

- Always wear rubber-soled boots when working near water and electrical systems.
- Install a Ground Fault Circuit Interrupter (GFCI) on electrical outputs to prevent shocks.
- Place warning markers around fast-moving or deep water setups to prevent accidental falls.
- Regularly inspect the structure after heavy rains or seasonal changes to catch any weaknesses.
- Avoid working alone near water flows; always have someone nearby in case of emergencies.
- Turn off all electrical systems before performing maintenance or adjustments.

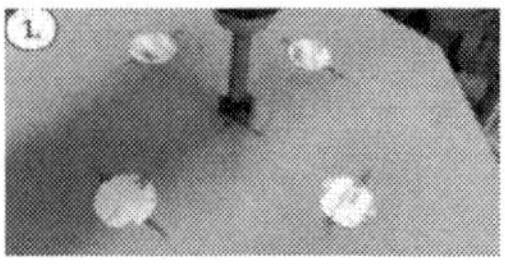

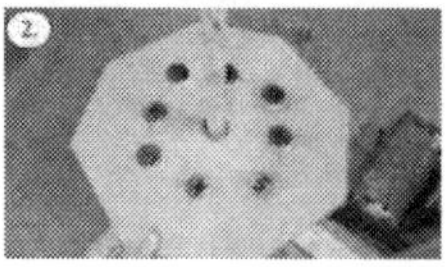

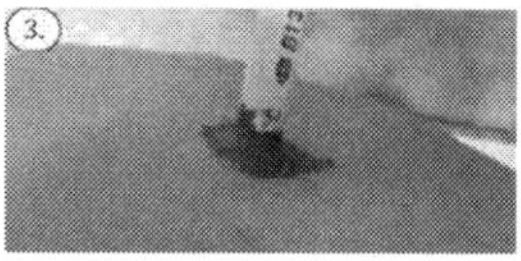

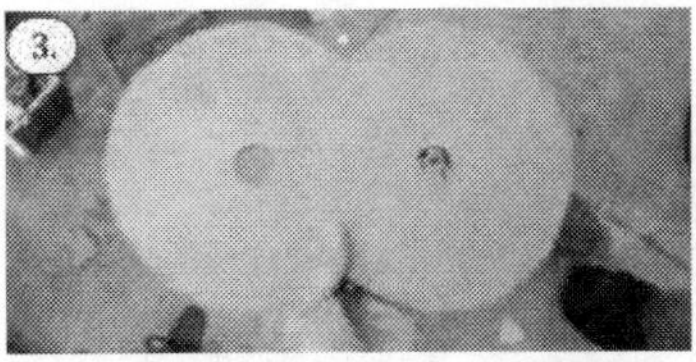

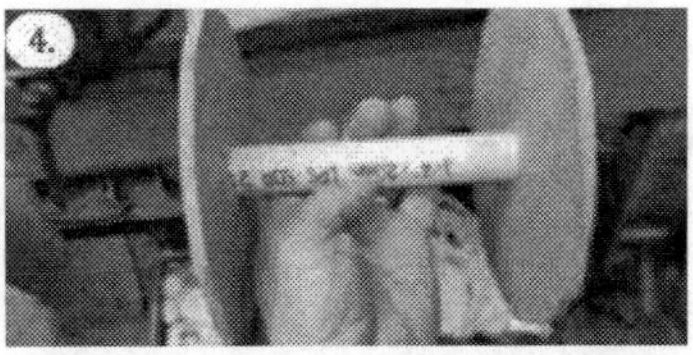

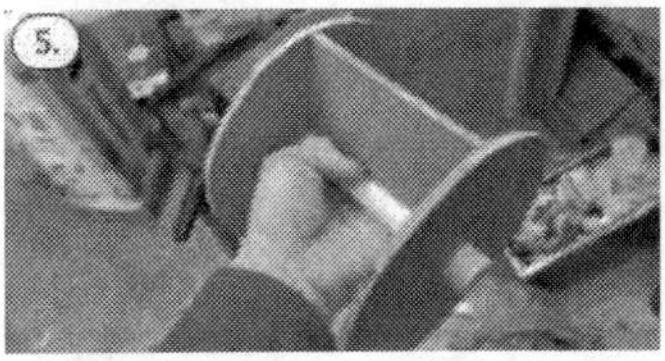

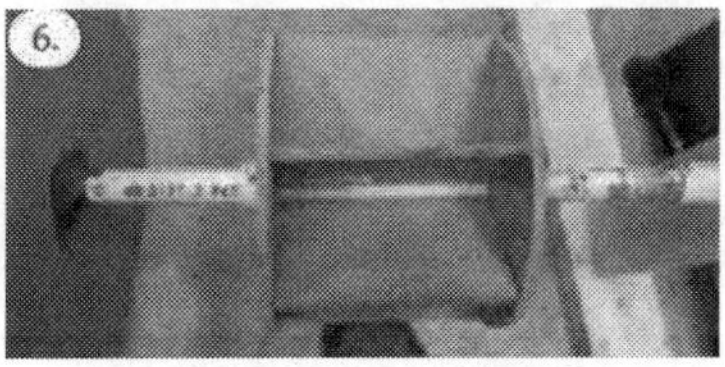

Integration Tip: Connect your hydroelectric output into your existing battery bank used by your solar panels or bicycle generator. By combining sources, you create a hybrid system that produces energy in all seasons and weather conditions.

Benefits:

- Provides continuous, renewable energy
- Works 24/7 without sunlight or human input
- Low maintenance once properly installed
- Highly scalable based on water availability

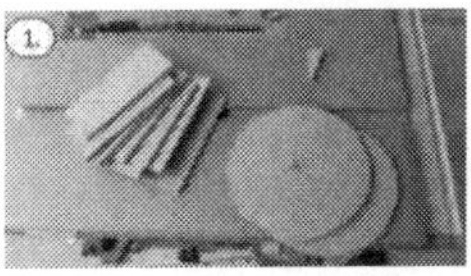

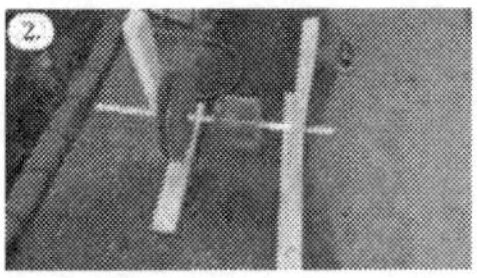

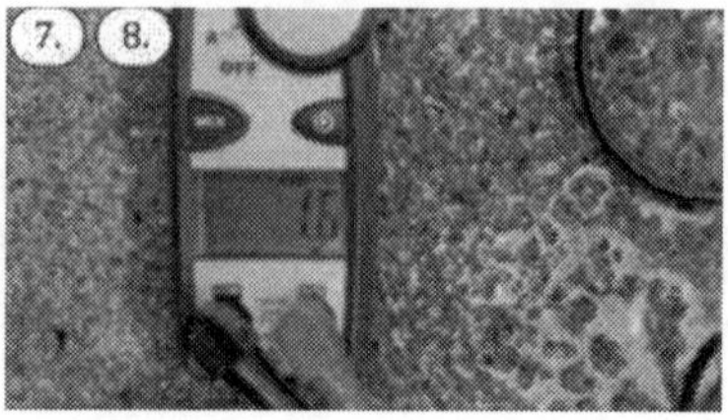

Sustainability Focus: Whenever possible, build waterwheels or turbine blades from salvaged wood, plastic barrels, or scrap metal. Reuse old motors from treadmills, washing machines, or electric scooters to keep project costs low and resource use minimal.

Chapter 10: Solar Power Solutions

The sun provides an endless supply of energy — free, abundant, and accessible almost anywhere. Capturing even a small fraction of this power can transform your off-grid property, providing electricity for lighting, refrigeration, water pumping, communication, and more.

Solar power is one of the most reliable renewable energy sources available. It scales easily from small portable panels to full home systems, requires minimal maintenance, and works well in tandem with other energy projects like hydroelectric and wind power.

This project will guide you through building a simple, expandable solar panel system, including choosing batteries, wiring safely, and positioning panels for maximum efficiency.

Materials Needed:

- Solar panels (100W to 400W recommended per panel)

- Deep-cycle batteries (Lead Acid or Lithium options)
- Charge controller
- Inverter (to convert DC to AC power if needed)
- Wiring (appropriately gauged)
- MC4 connectors and fuses
- Mounting hardware (metal rails, brackets, screws)
- Basic electrical tools (drill, screwdriver, multimeter, wrench set)

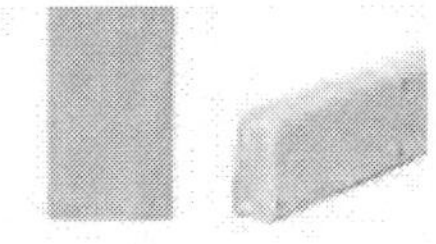

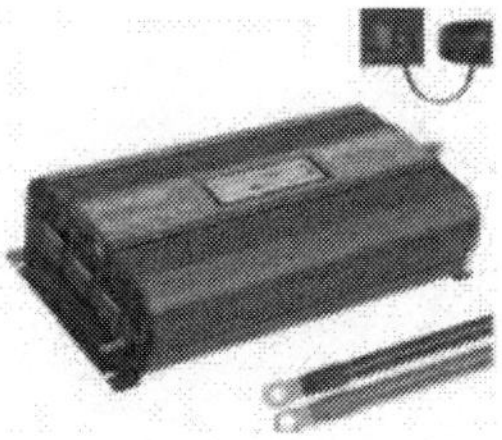

Building Your Own Solar Panel System:

1. **Select Your Batteries:**
    - **Lead Acid Batteries:** Affordable and reliable, but heavier and require ventilation.
    - **Lithium Batteries:** More expensive upfront but lighter, longer-lasting, and maintenance- free.
2. **Install the Panels:**
    - Mount panels on a south-facing roof or ground rack where they receive full sunlight.
    - Tilt panels at an angle equal to your local latitude for year-round performance.

3. **Wiring the System:**
    - Connect panels in series for higher voltage or parallel for higher amperage, depending on your system design.
    - Use MC4 connectors to simplify safe wiring.
    - Always install an inline fuse between panels and charge controller.
4. **Connecting Charge Controller:**
    - Wire the panels to the charge controller.
    - Wire the controller output to your battery bank.
5. **Adding the Inverter (Optional):**
    - Connect the inverter to the battery bank if you need AC power for home appliances.
    - Use heavy-gauge cables and proper fusing.
6. **Final Checks:**
    - Test voltage and amperage at each connection with a multimeter.
    - Secure all wiring and hardware against wind and weather exposure.

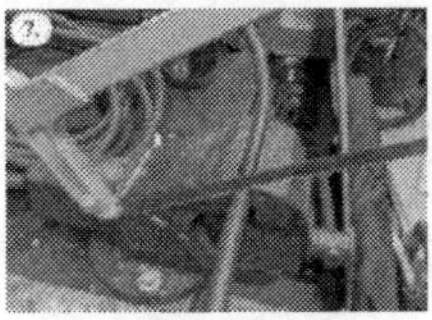

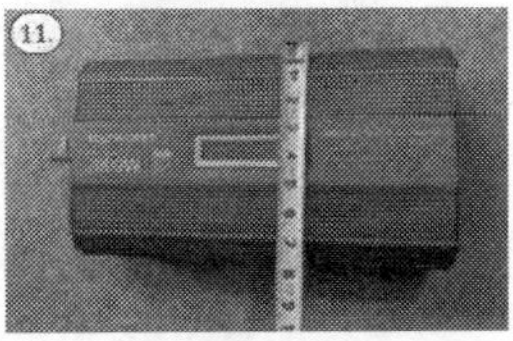

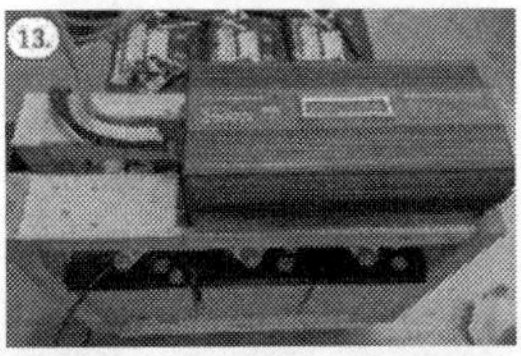

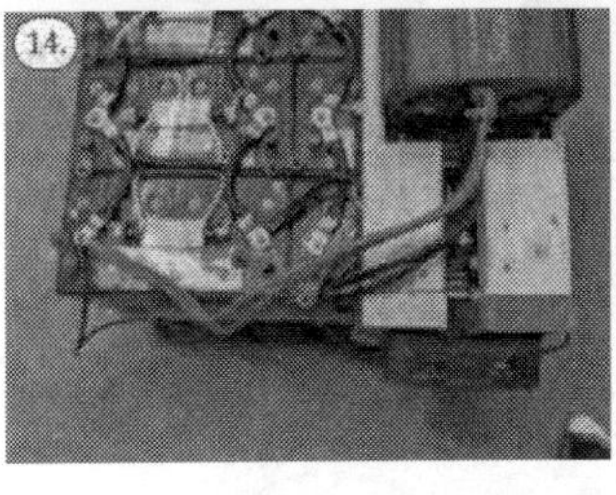

Designing and Installing the Battery Bank:

- Group batteries based on voltage needs (12V, 24V, or 48V systems).
- Use heavy-duty, insulated wiring.

- Install in a well-ventilated, temperature-stable area.
- Protect batteries from extreme heat, cold, and moisture.

Solar Panel Placement and Grid Connection:

- Avoid shaded areas, even partial shading can dramatically reduce output.
- Clean panels periodically to remove dust, pollen, and bird droppings.
- If grid-tied (optional for hybrid properties), install a proper inverter and work with a certified electrician.

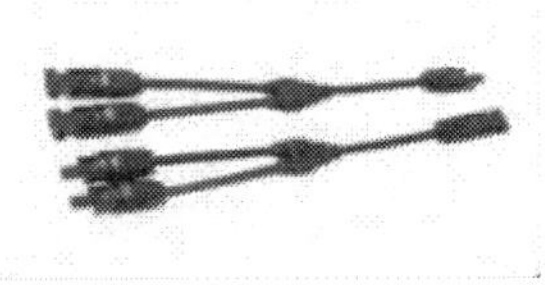

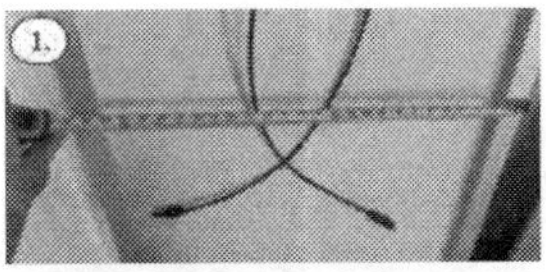
1.

3.

4.
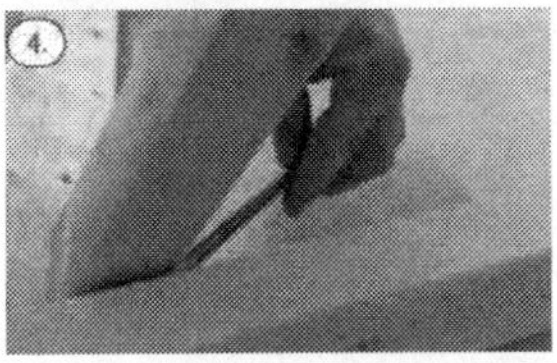
4.
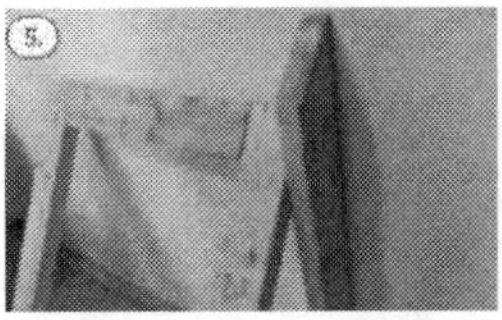
5.
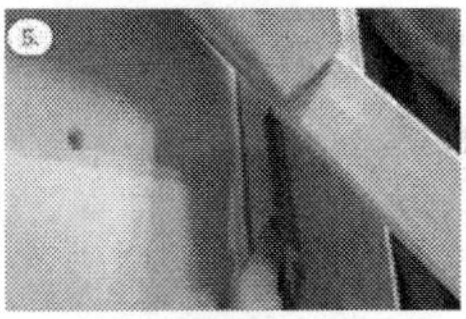
5.

7.

7.

8.

8.

1.
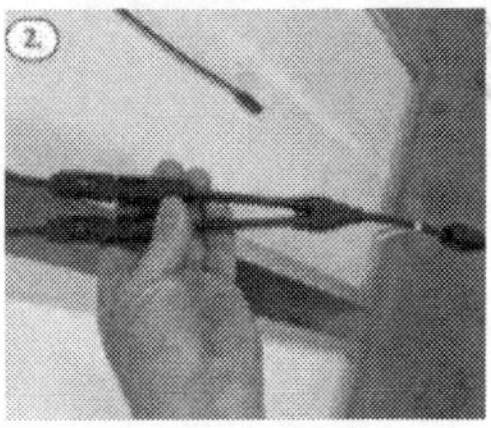
2.
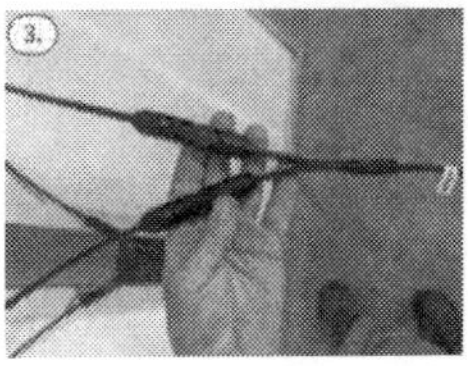
3.
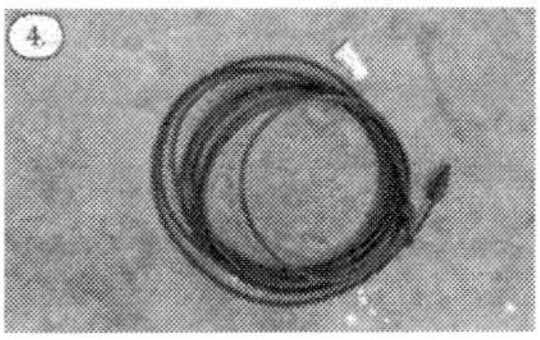
4.
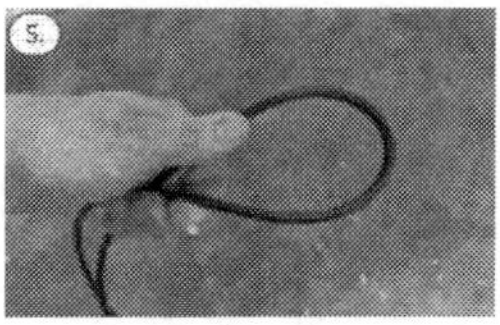
5.

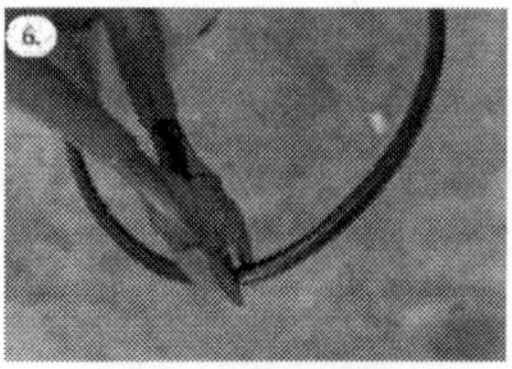

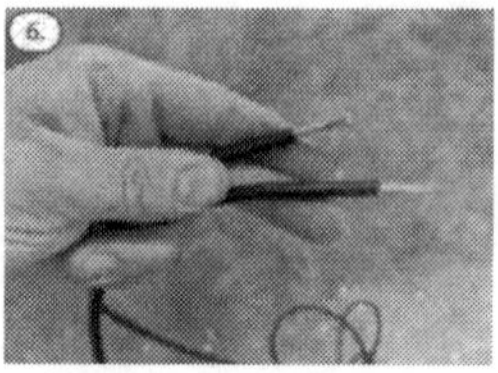

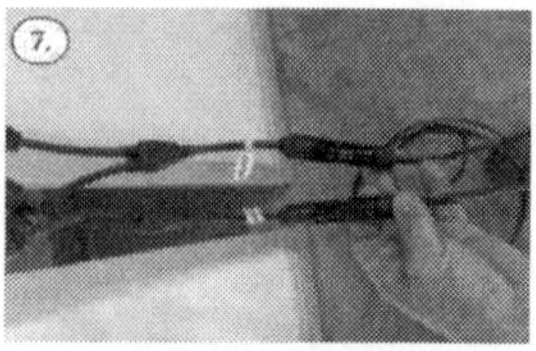

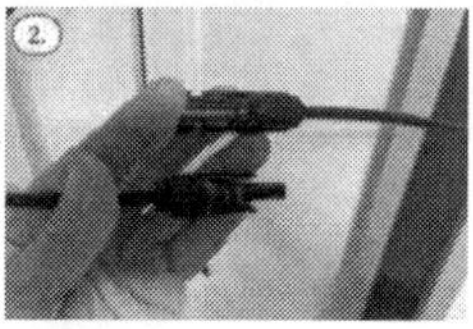

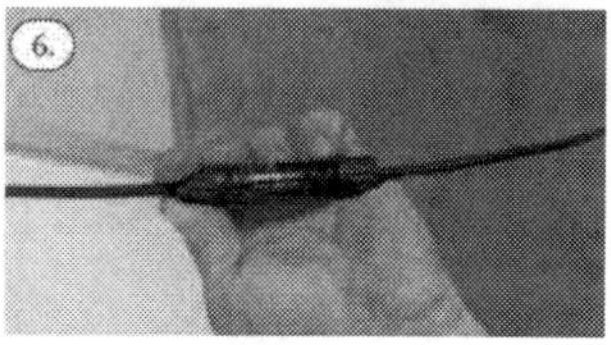

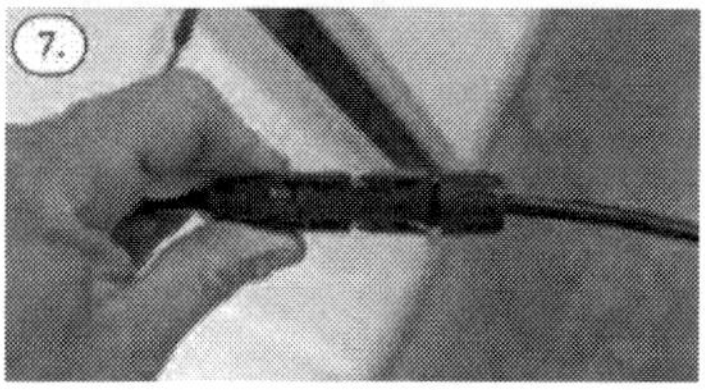

Safety Measures:

- Always disconnect panels from wiring before working on the system.
- Use insulated tools when handling electrical connections.
- Wear gloves and eye protection while wiring and mounting panels.
- Install all equipment according to manufacturer's specifications to avoid fire hazards.
- Never mix different battery types or panel types without appropriate controllers.

Integration Tip: Integrate your solar system with your hydroelectric and bicycle generator outputs. Use smart charge controllers that allow multiple energy inputs into a shared battery bank, ensuring you have reliable power regardless of weather conditions.

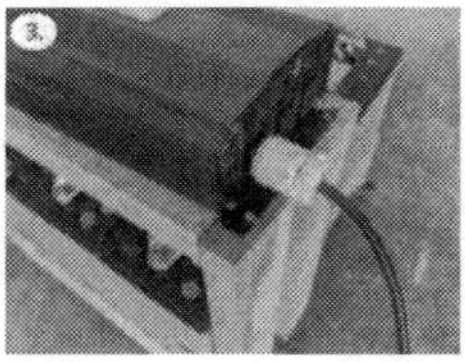

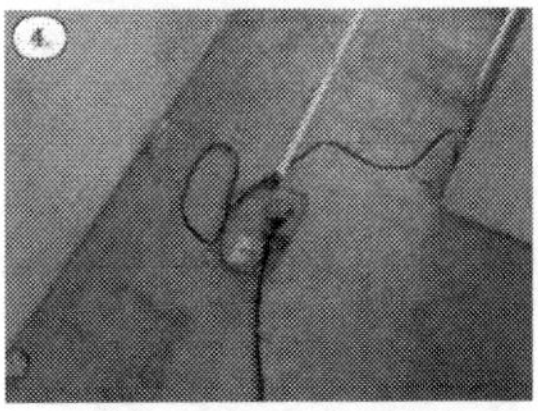

Benefits:

- Provides sustainable, renewable power
- Reduces or eliminates dependence on generators or fuel
- Low maintenance with high reliability
- Expandable to match growing energy needs

Sustainability Focus: Source panels, batteries, and inverters second-hand when possible. Many large solar farms upgrade equipment regularly, and used panels often have decades of service life remaining at a fraction of new costs. Reusing quality components reduces waste and supports sustainable building practices.

Chapter 11: How to Build Your Own Wind Turbine

Wind is a powerful and often underutilized energy source, especially in rural or open landscapes. Harnessing wind energy complements your solar, hydro, and manual energy systems, filling gaps during cloudy days, rainy seasons, or drought periods.

A homemade wind turbine not only captures another renewable energy stream but also reinforces the resilience and self-sufficiency of your property. By using salvaged materials and simple mechanical designs, you can build a functional wind generator at a fraction of the cost of commercial models.

When integrated with your solar panels, hydroelectric systems, and human-powered generators, a wind turbine completes the "energy net" — ensuring continuous power generation across different weather conditions.

Materials Needed:

- Permanent magnet DC motor (300-500W recommended)
- Blades (PVC pipe sections, wood, or repurposed metal)
- Tail vane (sheet metal or plywood)
- Tower pole (metal pipe, treated wood, or scaffolding)
- Guy wires and ground anchors
- Charge controller (wind-compatible)
- Rectifier (if motor outputs AC)
- Batteries and inverter (optional)
- Electrical wiring, connectors, and tools (drill, wrenches, saw)
- Multimeter

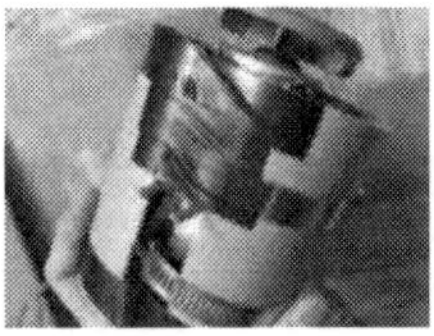
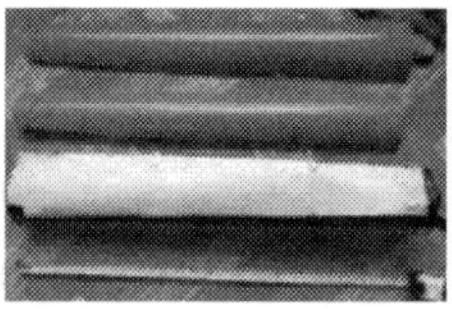

Tools Needed:

- Drill and drill bits
- Saw (hand saw, reciprocating saw, or jigsaw)
- Wrenches and sockets
- Screwdrivers

- Clamps
- Safety harness (for tower installation)

Building Process:

1. **Prepare the Blades:**
    - Cut PVC pipe or wood into three or more equal-length blades.
    - Shape the blades into an aerodynamic profile (one side rounded, one side flat) to improve efficiency.
2. **Assemble the Rotor:**
    - Mount the blades evenly around a

central hub (wheel hub or fabricated metal disc).
    - Bolt securely and balance the assembly to minimize vibration.
3. **Attach the Motor:**
    - Connect the rotor hub to the motor shaft.
    - Ensure tight, centered mounting to reduce wear and tear.
4. **Build the Tail Vane:**
    - Cut a sturdy piece of sheet metal or plywood.
    - Mount it behind the rotor to keep the turbine facing into the wind.

5. **Construct the Tower:**
    - Use a strong, tall pole to lift the turbine into stronger wind currents.
    - Anchor the tower with guy wires and ground stakes.
    - If possible, use tilt-up tower designs for easier maintenance.
6. **Wiring and Electrical Setup:**
    - Connect the motor output through a rectifier (if needed) to convert AC to DC.
    - Wire into a charge controller, then into

your battery bank.

- Install appropriate fuses and breakers for system protection.

7. **Testing and Balancing:**
   - Spin the blades by hand and check output with a multi-meter.
   - Make final adjustments to blade angle or tower alignment.

Safety Measures:

- Always install guy wires to stabilize the tower, especially in high-wind areas.

- Use a safety harness when assembling or servicing the turbine at height.
- Install lightning protection if feasible.
- Place warning markers around the base to prevent accidental collisions.
- Never approach the turbine during high winds or storms.

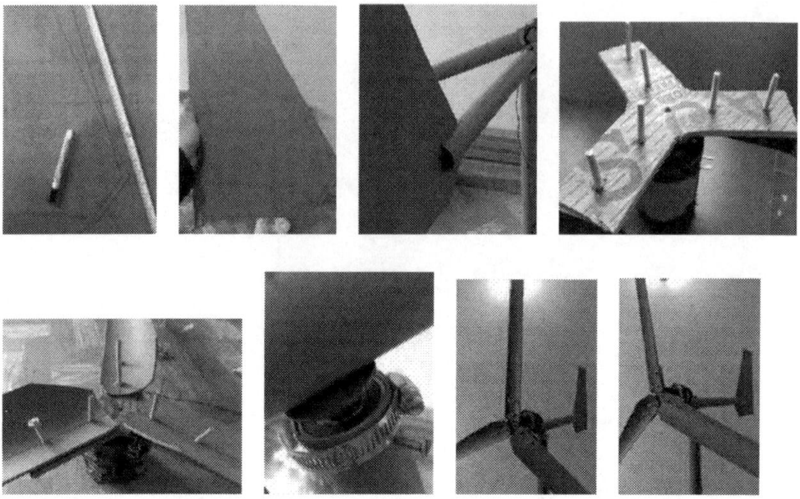

Integration Tip: Link your wind turbine into your existing solar and hydro battery banks using a hybrid charge controller. When the sun isn't shining and the rivers slow down, the wind often picks up — keeping your off-grid energy system balanced and resilient.

Use excess energy produced by wind to run non-essential systems like greenhouse fans, water heaters, or additional battery banks, ensuring nothing is wasted.

Benefits:

- Generates power during nights and storms
- Complements solar and hydro systems for full energy coverage

- Low maintenance with long operational life
- Expandable by adding more turbines over time

Sustainability Focus: Build blades from reclaimed PVC piping, old metal signage, or wooden boards. Salvage motors from treadmills, scooters, or industrial equipment to save costs and repurpose valuable components. Sourcing materials locally not only reduces expenses but strengthens your property's independence and environmental responsibility.

D. EMP & Communication Protection

In a truly resilient off-grid lifestyle, it isn't enough to simply produce food, water, and energy — you must also protect these systems from unseen threats.

An electromagnetic pulse (EMP), whether caused by natural solar flares or manmade events, can instantly disable electrical equipment, communication systems, and critical survival tools. Without preparation, an EMP could undo years of self-reliance in seconds.

Similarly, communication remains a lifeline in any crisis. Whether coordinating with neighbors, receiving emergency broadcasts, or alerting others to danger, maintaining robust communication channels is essential.

This section teaches you how to shield your home and equipment from EMP damage, repurpose everyday devices into powerful communication tools, and establish reliable off-grid networks.

By combining protection and communication projects, you create a layered safety net — ensuring your hard-earned independence survives even the most disruptive events.

Through careful design, use of local materials, and smart integration with your energy and security systems, you'll build a property that can stand strong when others falter.

Let's learn how to defend and connect.

Chapter 12:
How to Protect Your House Using an EMP Shield

Imagine building a water system, an energy network, and a food supply — only to lose them all in a flash. Electromagnetic pulses (EMPs) pose a real, often underestimated threat to modern technology and critical off-grid infrastructure.

Installing an EMP shield on your property is a crucial step toward holistic resilience. By protecting sensitive electronics, energy systems, and communication devices, you ensure your off-grid investment remains operational even during widespread disruption.

With a few simple steps and materials, you can dramatically increase your property's defenses against EMP attacks or natural solar storms.

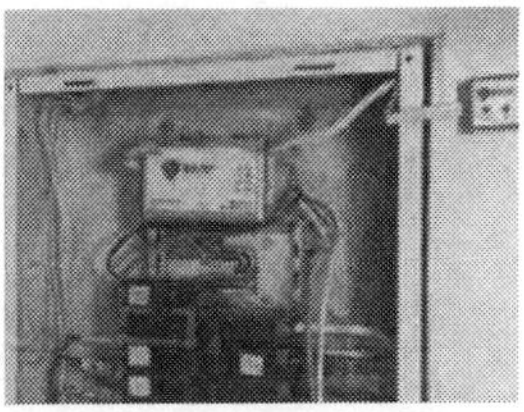

Materials Needed:

- EMP shield device (commercially available models or DIY Faraday cage setup)
- Grounding rod (copper or galvanized steel)
- Heavy-gauge copper grounding wire
- Metal enclosures (for smaller electronics)
- Surge protectors
- Basic electrical tools (drill, screwdrivers, wrench, multimeter)
- Safety gloves and glasses

Building Process:

1. **Select Your Shielding Strategy:**
    - **Commercial EMP Shields:** These devices attach to your home's main electrical panel, solar systems, or generators.
    - **DIY Faraday Cages:** For smaller items, you can create simple Faraday cages using galvanized steel trash cans, metal ammunition boxes, or even layered aluminum foil.

2. **Install the EMP Shield:**
    - Turn off all power at your main breaker.
    - Mount the EMP shield device according to manufacturer instructions, typically onto your breaker panel or near your solar inverter.
    - Connect the shield to a dedicated grounding rod using heavy copper wire.
3. **Grounding Installation:**
    - Drive the grounding rod deep into the earth (at least 6-8 feet).
    - Clamp the grounding wire securely to the rod and connect to your EMP shield.
    - Test grounding continuity with a multimeter.
4. **Protecting Individual Devices:**
    - Place critical electronics (radios, spare solar controllers, inverters) inside grounded Faraday cages when not in use.
    - Use quality surge protectors on all plugged- in devices.

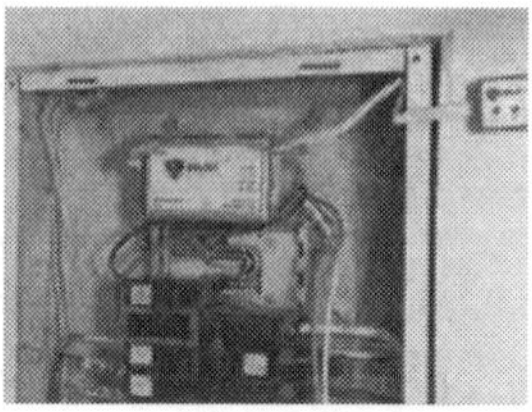

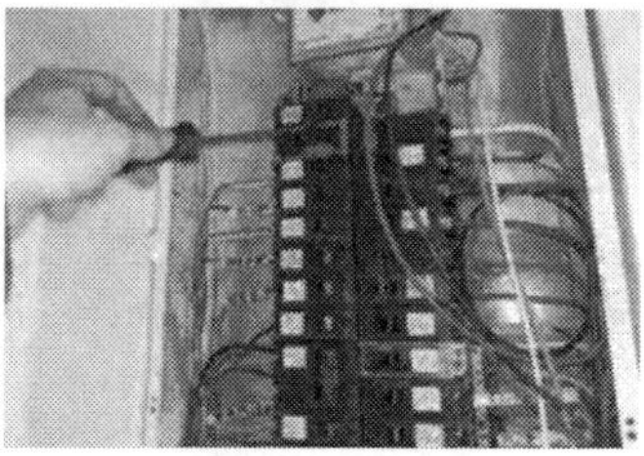
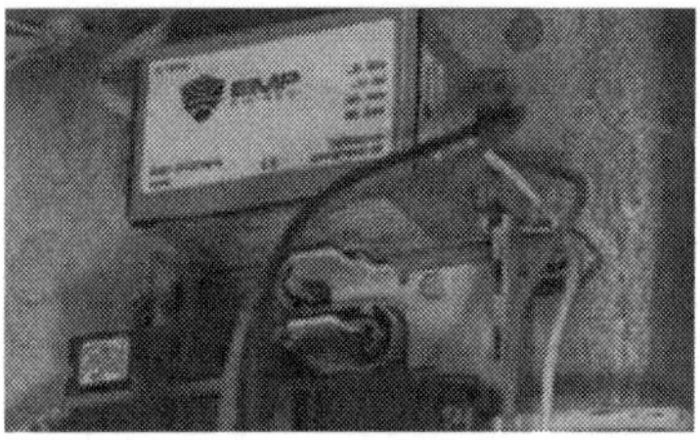

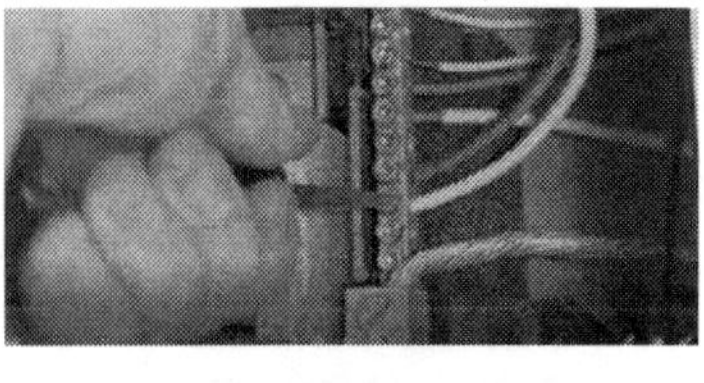
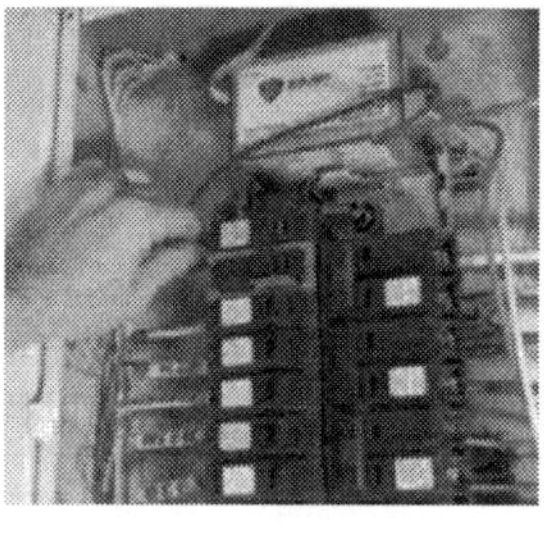

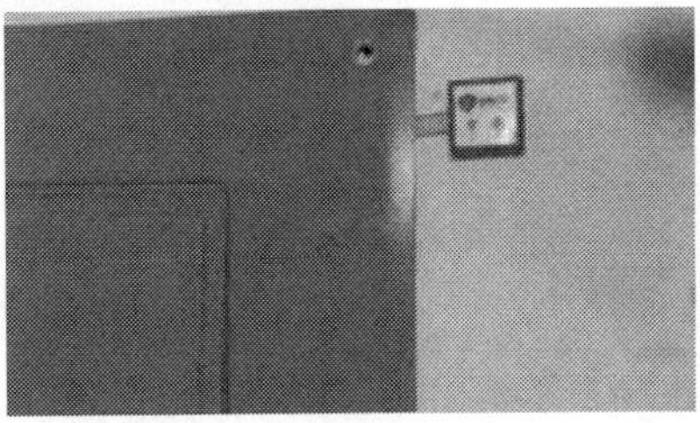

Safety Measures:

- Always disconnect all power sources before installation.

- Use insulated tools and wear safety gloves.
- Follow manufacturer grounding instructions carefully to avoid creating new electrical risks.
- Regularly inspect grounding connections and shield placement.

Integration Tip: Shield your entire energy system holistically – including solar arrays, wind turbines, hydro generators, and battery banks. Integrate small Faraday cages into your communication hubs to ensure radios and navigation equipment remain functional even after an EMP.

Benefits:

- Protects entire off-grid systems from catastrophic damage
- Extends life and reliability of expensive infrastructure
- Maintains critical communication and energy access during crises
- Provides peace of mind for long-term survival planning

Sustainability Focus: Repurpose metal containers, old safes, or ammunition boxes for DIY Faraday cages.

Reclaim copper grounding wire from surplus or demolition sites to reduce costs and environmental impact.

Chapter 13:
Turning a Car CB Radio into a Powerful Transmitter

In a grid-down situation, communication becomes your lifeline. A car CB (citizens band) radio, often overlooked in today's world of smartphones, can be a powerful tool for maintaining contact with neighbors, family members, or emergency groups over long distances.

With a few modifications, you can dramatically increase the range and effectiveness of a basic CB radio, turning it into a vital part of your off-grid communication network.

By combining simple upgrades with sustainable power sources from your solar, hydro, or wind systems, you create a resilient, always-on way to share information when it matters most.

Materials Needed:

- CB radio unit (12V car model)
- High-gain CB antenna (102” steel whip or fiberglass base antenna)
- Coaxial cable with connectors (RG-58 or RG-8 recommended)
- Mounting bracket or pole for antenna
- 12V battery or solar-powered system
- SWR meter (for tuning the antenna)
- Basic electrical tools (screwdrivers, pliers, wrenches)
- Electrical tape and zip ties

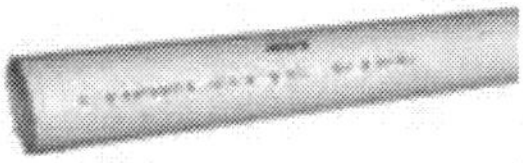

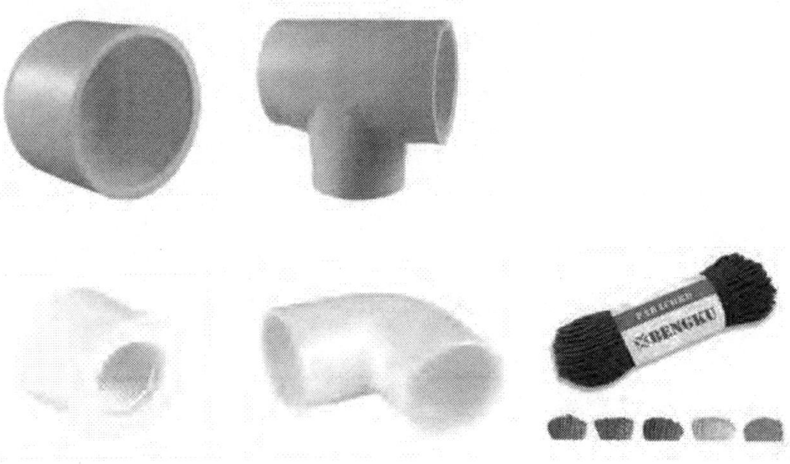

Building Process:

1. **Select the Right CB Radio:**
    - Choose a reliable 40-channel CB radio model designed for vehicle use.
    - Ensure it has an external antenna port.
2. **Install the High-Gain Antenna:**
    - Mount the antenna as high as possible: rooftops, towers, or poles work well.
    - Ensure the antenna is vertical and securely anchored.
    - Run coaxial cable from the antenna to your CB radio, avoiding sharp bends or

kinks.

3. **Tuning the Antenna:**
    - Use an SWR (Standing Wave Ratio) meter to fine-tune the antenna for best signal performance.
    - Follow SWR meter instructions carefully to avoid damaging the radio.
4. **Power Setup:**
    - Connect the CB radio to a 12V battery bank powered by your solar system or other off- grid energy source.
    - Install an inline fuse for safety.
5. **Testing and Adjustments:**
    - Power on the radio and perform test transmissions.
    - Adjust antenna placement and coax connections if needed to improve clarity.

CB Radio Reach Enhancement:

- Use longer, higher antennas to increase transmission distance.
- Operate on flat, open terrain to minimize signal obstructions.

- Keep coaxial cable runs as short as possible to reduce signal loss.

Building a CB Base Station and Antenna:

- Consider building a permanent base station in your home, complete with dedicated grounding rods and EMP protection.
- Integrate backup power options such as solar panels or bicycle generators to keep the station running during blackouts.

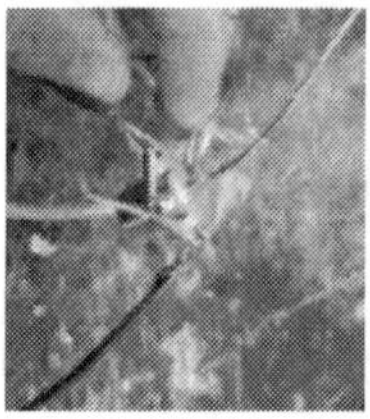

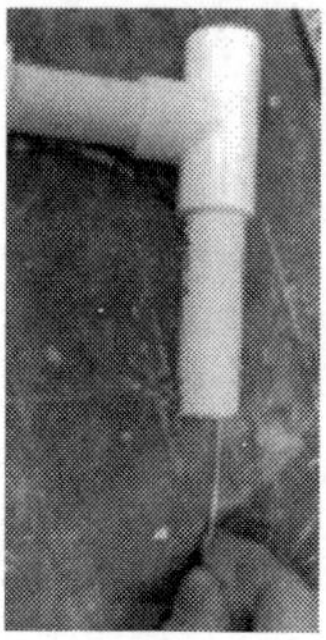
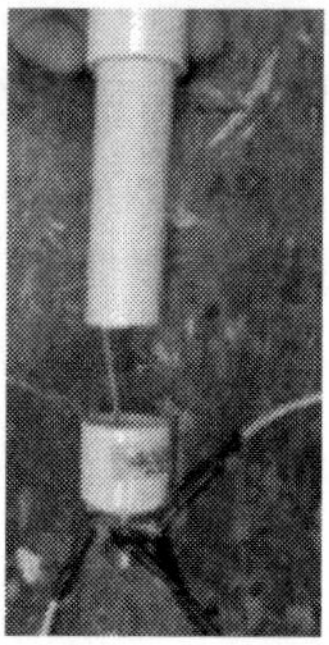
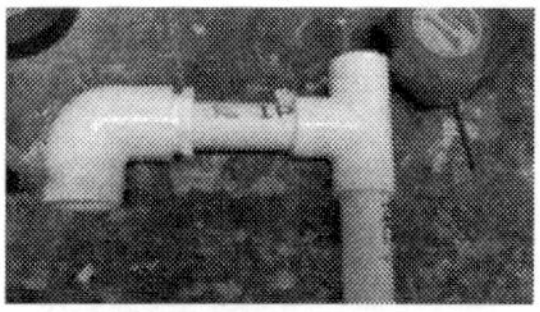

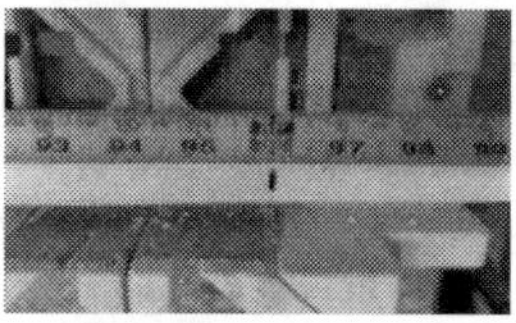

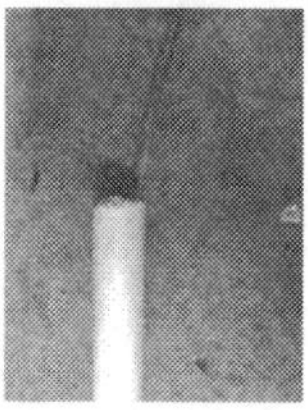
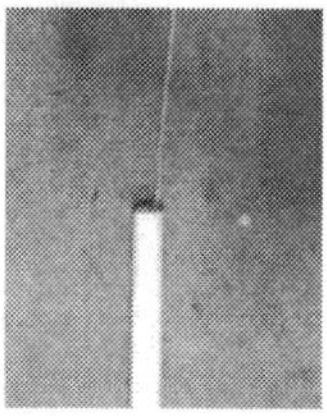

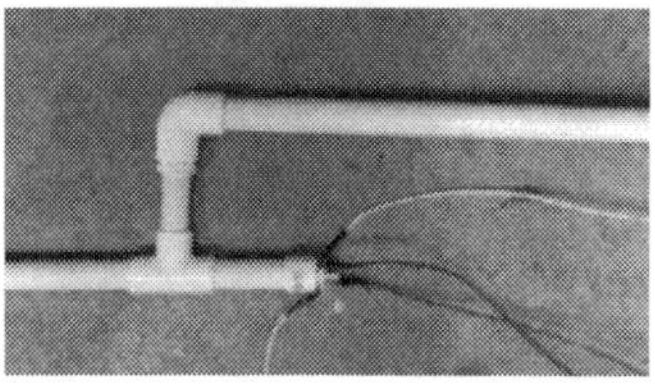
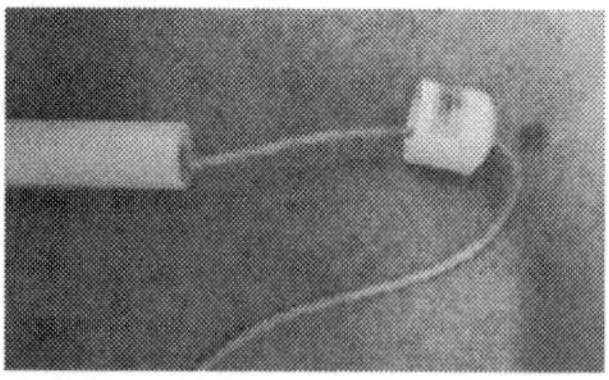

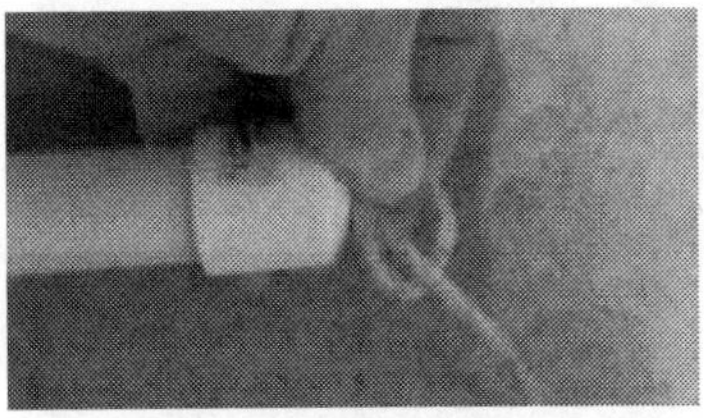

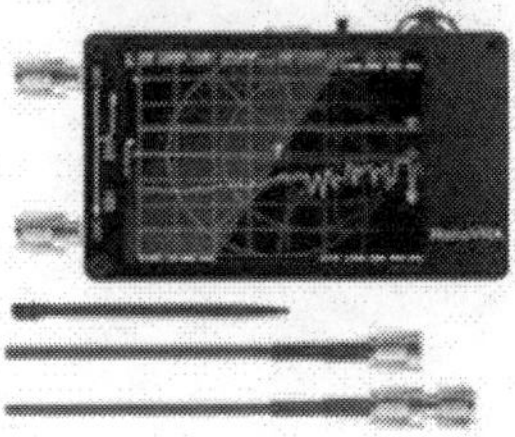
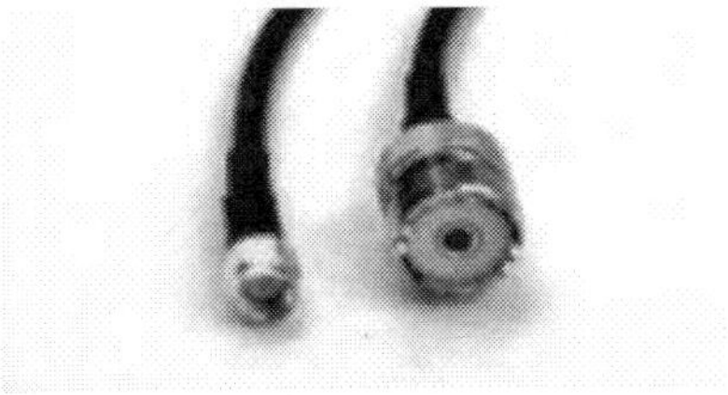
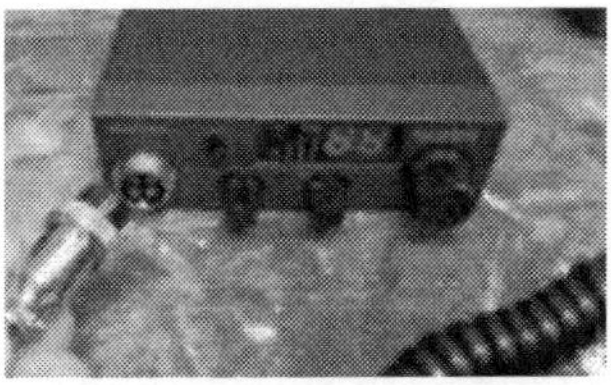

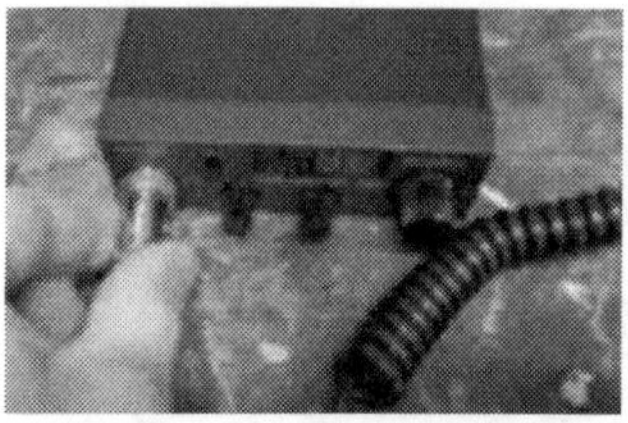
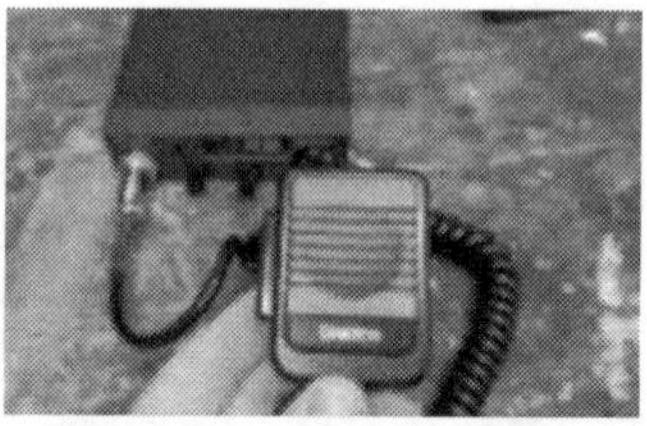
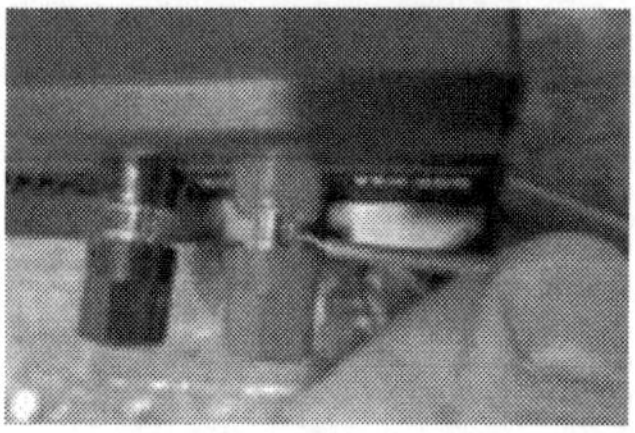

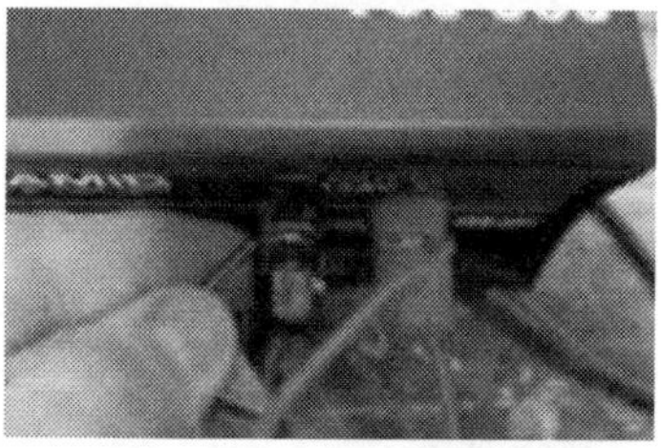
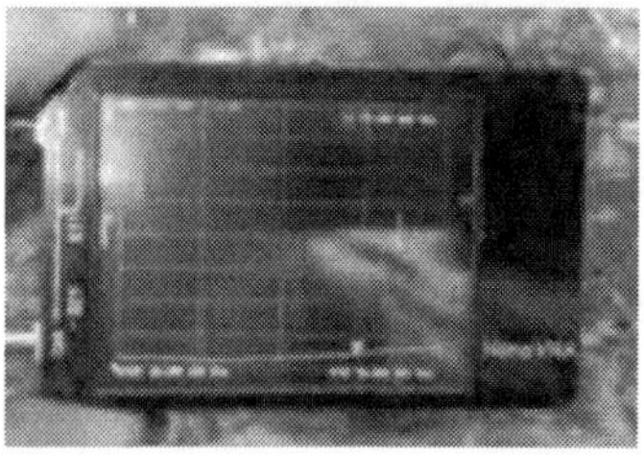

Safety Measures:

- Always ground your CB base antenna properly to prevent lightning damage.
- Avoid transmitting at maximum power for prolonged periods to reduce equipment

overheating.

- Ensure towers or high antenna poles are safely anchored and marked.

Integration Tip: Integrate your upgraded CB system into your broader off-grid communications network. Pair it with FRS, GMRS, or HAM radios for layered communication redundancy across different ranges and conditions.

Power the CB station through your solar battery banks, linking communication and energy systems into one resilient infrastructure.

Benefits:

- Provides reliable two-way communication during emergencies
- Extends range far beyond standard handheld radios
- Enhances neighborhood coordination and situational awareness
- Operates fully off-grid with solar or manual energy sources

Sustainability Focus: Repurpose old CB radios and salvage antennas from flea markets, junkyards, or surplus sales. Extend the life of quality vintage equipment and minimize new manufacturing impacts.

Chapter 14: How to Use a Ham Radio in a Blackout

When traditional communication networks collapse, ham radio remains one of the few reliable ways to stay connected with the outside world. Known for its incredible range and versatility, ham radio can bridge vast distances, helping off-grid families coordinate, gather critical information, and maintain vital links to nearby communities.

With basic licensing, the right equipment, and integration into your renewable energy systems, ham radio becomes the backbone of a resilient communication plan — ready to operate long after cell towers and internet lines fall silent.

Materials Needed:

- Handheld (HT), mobile, or base-station ham radio
- High-gain antenna (dual band for VHF/UHF)
- Coaxial cable and connectors
- SWR meter (for antenna tuning)
- 12V battery bank or solar generator
- License (Technician class or higher)
- Programming cable and software (optional but helpful)
- Backup communication plan (frequency lists, call signs)

Building Process:

1. **Licensing and Preparation:**
    - Obtain at least a Technician class ham radio license. Online courses and practice exams are widely available.
    - Learn basic radio etiquette and emergency communication protocols.
2. **Select and Set Up Your Radio:**
    - Choose a durable, reliable ham radio model.
    - For base stations, mount a permanent high- gain antenna as high as feasible.
    - Use portable HT (handheld) units for mobile or backup use.
3. **Power Your System:**
    - Connect your radio to a solar-charged battery bank for full off-grid capability.
    - Install inline fuses to protect both radio and battery.
4. **Antenna Tuning:**
    - Use an SWR meter to fine-tune your antenna for maximum signal clarity and safe operation.
    - Ensure coaxial cable runs are clean

and weatherproofed.

5. **Programming Channels:**
    - Program local repeaters, emergency frequencies, and family call signs into your radio memory.
    - Include simplex channels for direct radio-to- radio communication when repeaters fail.

Equipment and Setup Considerations:

- **Handheld Radios (HT):** Compact, versatile, excellent for field use.
- **Mobile Radios:** Higher power output (up to 50W), better for vehicles or semi-portable bases.
- **Base Stations:** Highest power and best range when combined with proper antenna installations.

Emergency Usage Tips:

- Keep laminated frequency lists and call procedures ready.
- Conduct monthly communication tests with family and neighbors.
- Monitor weather and emergency services

broadcasts even during "normal" times.

Safety Measures:

- Properly ground all external antennas to protect against lightning.
- Operate at legal power levels to avoid overheating or equipment damage.
- Always practice safe tower climbing procedures if installing large antennas.

Integration Tip: Combine your ham radio network with your upgraded CB system, walkie-talkies (FRS/GMRS), and even signal mirrors or flags for a full-spectrum communication setup. Power your radios directly from your renewable energy grid, ensuring continuous operation even in long-term outages.

Benefits:

- Provides worldwide communication ability
- Remains operational during widespread blackouts
- Strengthens neighborhood and mutual aid group coordination
- Connects to emergency networks, news, and rescue operations

Sustainability Focus: Buy used radios and antennas from trusted ham operators or swap meets. Many "retired" gear sets remain fully operational for decades with minimal maintenance, offering high-quality performance at a fraction of new equipment costs.

II. Defense, Ammunition, & Security Projects

A. Ammunition Storage & Firearm Security

Protecting your resources includes safeguarding your ability to defend your property, family, and systems. In a true off-grid survival situation, responsible ammunition storage and firearm security become critical.

Properly stored ammunition ensures you have a reliable supply when you need it most – whether for hunting, protection, or bartering. Firearm security, including hidden storage solutions, prevents theft, unauthorized access, and preserves your readiness without advertising your assets to the world.

This section covers long-term ammo preservation, covert gun storage builds, and strategic planning to integrate defensive measures into your overall self-reliance strategy.

As with every system in this guide, these projects prioritize local sourcing, sustainable methods, integration into broader systems, and layering security for redundancy.

Let's start by ensuring your ammunition supply is protected for the long haul.

Chapter 15: Ammo for Long-Term Storage

In off-grid living, every bullet counts. Whether used for hunting, defense, or trade, ammunition is a critical asset that must be protected from the elements, corrosion, and time.

Without proper storage, ammo can become unreliable or even dangerous. But with careful preparation, you can ensure your stockpile remains safe, effective, and ready for use years or even decades into the future.

This project guides you through building a sustainable ammunition storage system using locally available materials, smart preservation techniques, and strategic integration with your overall off-grid defense plan.

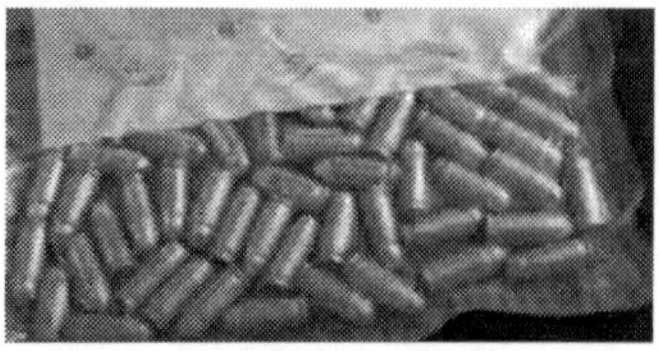

Materials Needed:

- Military surplus ammo cans (with rubber gaskets)
- Silica gel packs or moisture absorbers
- Vacuum sealer (optional)
- Desiccant packs
- Oxygen absorbers

- Heavy-duty plastic zip bags or Mylar bags
- Rust inhibitor paper (VPI paper)
- Labeling materials (permanent marker, waterproof labels)
- Storage shelving or cabinets

Building Process:

1. **Inspect and Prepare Containers:**
   - Ensure ammo cans or storage boxes have intact rubber gaskets and no rust.
   - Clean and dry the inside thoroughly.
2. **Moisture Control:**

- Place silica gel packets or desiccant packs inside each container.
- Add oxygen absorbers if you live in a humid region.

3. **Sealing Ammunition:**
    - For ultimate preservation, vacuum seal ammo boxes or use Mylar bags.
    - Double-bagging inside plastic zip bags provides added protection.
4. **Rust Prevention:**
    - Line the inside of the container with VPI paper to inhibit corrosion.
5. **Label and Organize:**
    - Clearly label containers with caliber, quantity, and date of storage.
    - Rotate older stock to the front for first use (First In, First Out system).
6. **Storage Location:**
    - Store ammo containers off the ground on shelving units.
    - Choose cool, dry, and dark areas such as basements, root cellars, or hidden compartments.

Protecting Ammo and Additional Storage Practices:

- Periodically inspect ammo cans for gasket degradation or moisture intrusion.
- Recondition desiccant packs by drying them in a low-temperature oven when saturated.
- Keep a manual inventory log (not just digital) to track stock levels.

Safety Measures:

- Never store ammunition near fuel, volatile chemicals, or extreme heat sources.
- Keep ammo separate from loaded firearms to reduce accidental discharge risk.
- Use childproof locks or hidden compartments if children are present.

Integration Tip: Integrate your ammunition storage near your hidden gun storage shelves (covered in the next project) to create a centralized, covert defense hub. Connect storage to your broader off-grid defense plan, pairing it with surveillance, perimeter defenses, and communication systems.

Benefits:

- Preserves ammunition reliability and lifespan
- Reduces the risk of corrosion and moisture damage
- Organizes supplies for quick access in emergencies
- Enhances property defense readiness without drawing attention

Sustainability Focus: Reclaim military surplus ammo cans or repurpose airtight storage containers from food services or industrial suppliers. Buying used containers reduces waste, lowers costs, and often provides higher durability than new plastic options.

Chapter 16:

Hidden Gun Storage Shelf & Covert Firearm Shelving

In a true off-grid setup, discretion is often the first line of defense. Securing firearms in hidden, covert storage locations allows you to maintain readiness without drawing unwanted attention to your armaments.

Rather than traditional safes that scream "valuables inside," a hidden gun storage shelf or covert shelving system blends seamlessly into your living environment. With smart design and careful material selection, you can create functional, camouflaged storage spaces that protect your tools while preserving the natural aesthetics of your off-grid home.

This project teaches you how to build a hidden shelf

with a locking mechanism, using sustainable materials and simple carpentry techniques accessible to any skill level.

Materials Needed:

- Wooden board (reclaimed hardwood, pallet wood, or pine)
- Magnetic locks or RFID locking mechanism
- Concealed hinges
- Gas struts or hydraulic arms (optional for easy opening)
- Screws, wood glue, nails
- Paint, stain, or finish to match existing decor
- Basic tools: saw, drill, screwdriver, level, measuring tape
- Felt, foam, or padded liner (optional)
- Anchoring hardware for wall mounting

Building Process:

1. **Design Your Shelf:**
    - Measure the space where you plan to mount the shelf.
    - Choose dimensions that balance storage capacity with inconspicuous appearance.

2. **Construct the Shelf Box:**
   - Cut and assemble wood panels into a hollow box.
   - Secure with wood glue and screws for maximum strength.
3. **Install Hinges:**
   - Mount concealed hinges along one side of the shelf bottom to create a hidden drop- down door.
   - Use gas struts for smooth, silent opening if desired.
4. **Install Locking Mechanism:**
   - Install magnetic locks or an RFID locking system inside the shelf.
   - Hide the activation point beneath a decorative element (small trim piece, faux drawer handle, etc.).
5. **Finishing Touches:**
   - Stain, paint, or finish the shelf to match surrounding furniture.
   - Line the inside with felt, foam, or padding to protect firearms.
6. **Mounting the Shelf:**
   - Secure the shelf to wall studs using heavy- duty anchors.

- Test opening and closing mechanisms thoroughly.

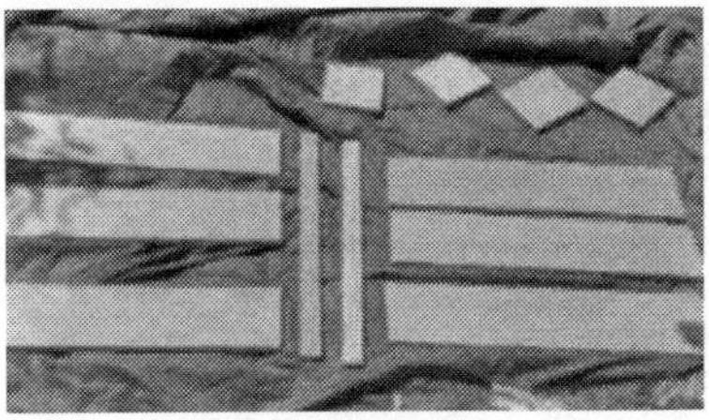
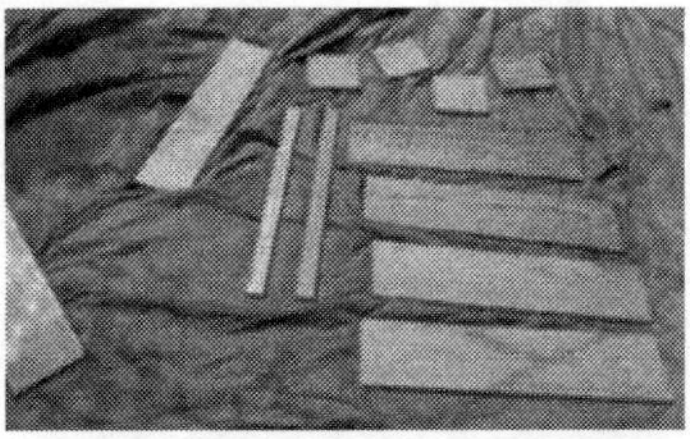

Placement Variations:

- Above doors
- Inside closets
- Integrated into bookshelves
- Behind mirrors or artwork

Safety Measures:

- Always install childproof secondary locks if

minors are present.

- Train household members on how to safely and discreetly access firearms.
- Regularly inspect locking systems for wear and function.

Integration Tip: Pair hidden firearm shelving with your ammunition storage setup to create a centralized, low- profile "defense node." Integrate security cameras, motion detectors, or tripwire alarms nearby to further harden your critical access points.

Benefits:

- Conceals firearms in plain sight
- Maintains rapid access in emergencies
- Protects weapons from theft and unauthorized access

- Blends seamlessly with off-grid, natural home aesthetics

Sustainability Focus: Repurpose reclaimed wood from demolition sites, barn salvage, or local scrap lumber yards. Salvage hinges and mounting hardware from old furniture whenever possible. Creative reuse reduces material costs and minimizes environmental impact.

Chapter 17: Ghost Guns

Built privately without serial numbers — offer an empowering option for off-grid defense planning.

By fabricating your own firearms using 80% lower receivers and In a future where regulations tighten and supply chains falter, the ability to build and maintain your own firearms represents a critical level of self-reliance. "Ghost guns" — firearms DIY kits, you not only maintain autonomy but also develop valuable mechanical skills that can serve across many aspects of off-grid living.

This project explores what ghost guns are, how to responsibly and legally build them, and how to integrate them thoughtfully into your broader security framework.

What is a "Ghost Gun"?

- A "ghost gun" is typically a firearm assembled by an individual from parts, especially starting with an unfinished (80%) lower receiver.

- 80% lowers are incomplete firearm frames or receivers that require additional machining to become functional.
- Because they are unfinished, they are not classified as firearms under federal law (though state laws vary — always verify local regulations).

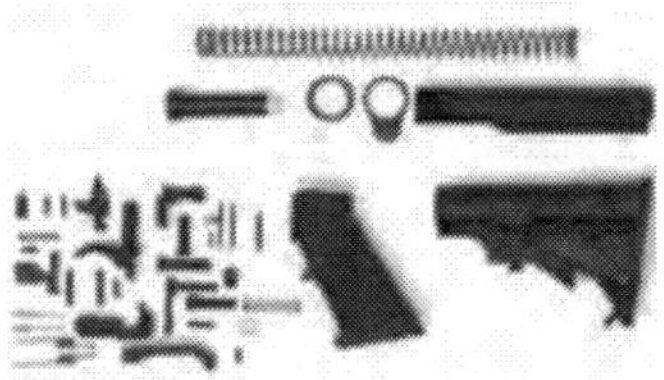
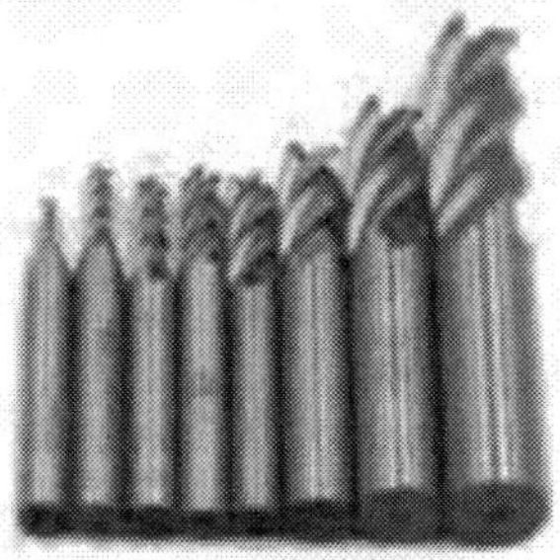

Materials Needed:

- 80% lower receiver (AR-15, Glock, or other platform)
- 80% lower jig kit

- Drill press or router (depending on jig type)
- Drill bits and end mills (provided with jig kit)
- Upper receiver and parts kit (bolt, barrel, trigger assembly, etc.)
- Punch set, vise, and basic hand tools
- Safety glasses and gloves
- Instruction manual (included with quality kits)

Building a Ghost Gun: Step-by-Step Guide:

1. **Research and Plan:**
   - Select your firearm platform based on your defensive needs (rifle, pistol, etc.).

   - Purchase an 80% lower and a matching jig kit from a reputable supplier.
2. **Set Up Your Workspace:**
   - Create a clean, well-lit, and ventilated workspace.
   - Secure the lower receiver firmly into the jig.
3. **Drill and Mill the Receiver:**
   - Following the jig instructions, drill pilot holes at designated points.
   - Mill out the fire control pocket to

proper depth.
  - Carefully drill the trigger, hammer, and selector holes.

4. **Deburr and Clean:**
   - Smooth any sharp edges with fine files or sandpaper.
   - Thoroughly clean the receiver to remove metal shavings and dust.
5. **Assemble the Firearm:**
   - Install the lower parts kit: trigger group, magazine catch, safety selector, etc.
   - Attach the upper receiver (barrel and bolt assembly) to complete the firearm.
6. **Function Check:**
   - Conduct a full mechanical safety check before loading any ammunition.
   - Dry fire carefully to ensure proper trigger reset and operation.

Safety Measures:

- Always wear eye protection when drilling and milling.

- Double-check jig settings before making each cut.
- Use sharp, undamaged drill bits and end mills to ensure clean, accurate cuts.
- Perform careful final inspections to verify mechanical safety before use.

Legal Considerations:

- Verify all local, state, and federal laws before building.
- As of writing, federal law allows building a firearm for personal use but prohibits selling without proper licensing.
- Some jurisdictions require serialization or have additional restrictions.

Integration Tip: Ghost guns provide a strategic advantage when integrated into your broader security system. Secure completed builds in your hidden gun storage areas.
Maintain a parts inventory for repairs or future builds, and consider creating multiple platforms (handgun and rifle) for layered defensive capabilities.

Benefits:

- Full control over the creation and maintenance of defensive tools

- Eliminates reliance on retail supply chains
- Teaches valuable mechanical, machining, and gunsmithing skills
- Enhances off-grid independence and resilience

Sustainability Focus: Whenever possible, source durable, quality parts that can withstand long-term off-grid use.

Salvage usable components from damaged firearms or surplus parts suppliers. Building "from scratch" encourages material conservation and mechanical self-sufficiency.

B. Black Powder, Reloading, & Ammunition Recycling

Off-grid living demands not just stockpiling resources, but mastering the skills to replenish them. In a prolonged crisis, traditional ammunition supplies can dry up — but with basic materials and know-how, you can produce your own black powder, reload spent cartridges, and recycle valuable components.

Learning to make black powder and reload ammo provides a tremendous advantage: it ensures you can continue hunting, defending your homestead, and bartering for supplies long after manufactured options disappear.

This section will guide you through simple, effective methods for producing and recycling essential ammunition supplies, using a mix of DIY ingenuity and sustainable resource management.

Let's begin with one of the oldest and most powerful self- sufficiency skills: making black powder.

Chapter 18: How to Make Black Powder

Black powder, also known as gunpowder, has powered firearms, cannons, and fireworks for centuries. For an off- grid homesteader, learning to make black powder ensures you have a reliable source of propellant for muzzleloaders, homemade ammunition, and even small-scale blasting tasks.

While commercially available black powder is highly regulated, producing it on a small scale for personal use is achievable with careful attention to safety and quality.

This project will teach you two methods — dry mixing and wet mixing — to produce functional, safe black powder at home.

Materials Needed:

- Potassium nitrate (saltpeter)
- Charcoal (homemade or bought)
- Sulfur powder
- Mortar and pestle or ball mill (for grinding)
- Measuring scale
- Fine mesh sieve
- Distilled water (for wet method)
- Non-sparking mixing tools (wood, plastic)
- Heavy rubber gloves
- Eye protection and dust mask
- Storage containers (sealed, non-metallic)

Proportions for Black Powder:

- 75% potassium nitrate
- 15% charcoal
- 10% sulfur

Dry Mixing Method:

1. **Grind the Ingredients:**
   - Using a ball mill or mortar and pestle, grind each ingredient separately into fine powders.
   - Be patient and gentle; excessive friction

can cause ignition.

2. **Combine Carefully:**
    - Weigh the ingredients precisely.
    - Mix them together using a plastic spoon and a plastic or wooden bowl.

    - Use a gentle folding motion — no grinding or forceful stirring.
3. **Sieve the Mixture:**
    - Pass the combined powder through a fine mesh sieve to ensure even mixing.
4. **Dry and Store:**
    - Store the dry black powder in a cool, dry, sealed non-metallic container.

Wet Mixing Method (Preferred for Safety):

1. **Create a Slurry:**
    - After grinding separately, mix the powders with a small amount of distilled water to form a paste.
2. **Knead Gently:**
    - Press and knead the paste using a wooden spoon or plastic scraper.
3. **Form into Cakes:**
    - Spread the mixture out on a flat

surface (lined with wax paper) and press into thin, uniform cakes.

4. **Dry Completely:**
   - Air dry the cakes thoroughly (can take several days).

5. **Crush and Sieve:**
   - Once dry, gently crush the cakes into fine grains and sieve to desired consistency.

Dry vs. Wet Methods:

- **Dry Mixing:** Faster but riskier due to static discharge.
- **Wet Mixing:** Safer and produces more consistent grains.

Safety Procedures:

- Always work outdoors, away from flames, sparks, or static electricity.
- Wear protective gloves, dust masks, and goggles.
- Never store large amounts of black powder in one container.
- Never grind mixtures — only individual ingredients.

Tools and Testing:

- Use a small test batch to verify burn rate and consistency.
- Adjust charcoal amount for faster or slower burns.
- Light test batches using a long stick or remote ignition method.

Integration Tip: Pair your black powder production with reloading skills and black powder firearm maintenance.

Establish a "powder pantry" alongside your ammunition stores, ensuring layered defensive capabilities.

Benefits:

- Provides renewable source of firearm and tool propellant
- Reduces dependence on commercial ammunition suppliers
- Strengthens self-sufficiency and emergency resilience
- Teaches valuable historical survival skills

Sustainability Focus: Make your own charcoal from hardwood scraps like willow, alder, or maple. Collect sulfur from local mineral deposits if possible, and harvest potassium nitrate from aged manure piles or compost heaps with the right processing techniques.

Chapter 19: Reloading Processes

Reloading ammunition is one of the most empowering skills a self-reliant individual can master. It transforms spent casings into fully functional rounds, drastically reducing dependency on outside suppliers while saving money and conserving resources.

Whether you're hunting, defending your property, or preparing for long-term survival, reloading allows you to maintain a consistent and reliable ammunition supply even when store shelves are bare.

This project covers the complete reloading cycle, from setting up your workspace to producing safe, consistent ammunition using simple tools and sustainable methods.

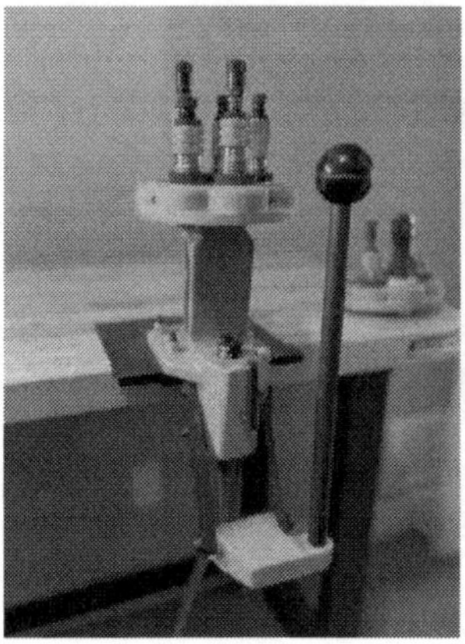 

Materials Needed:

- Reloading press (single-stage press recommended for beginners)
- Reloading dies (specific to your caliber)
- Spent brass casings (cleaned and inspected)
- Primers
- Gunpowder
- Bullets (commercial or cast at home)
- Case lube
- Calipers (for precise measurements)
- Reloading manual (critical for safe powder loads)
- Safety glasses and gloves

Installing Your Press:

1. **Choose Your Workspace:**
   - Select a sturdy, well-lit workbench or

heavy table.

  - Ensure the surface can handle repetitive pressure and vibration.

2. **Mount the Reloading Press:**
   - Bolt the press securely to the table using heavy-duty bolts.
   - Check for stability — the press should not shift during operation.

Reloading Dies:

- **Sizing Die:** Resizes spent brass back to factory dimensions.

- **Seating Die:** Seats the bullet into the case to the proper depth.
- **Crimp Die (optional):** Applies a slight crimp to hold the bullet firmly in place.

Reloading Steps:

1. **Case Preparation:**
   - Clean spent brass (use tumbler, ultrasonic cleaner, or hand-cleaning).
   - Inspect for cracks, splits, or deformities — discard any damaged casings.
2. **Sizing the Case:**

- Apply case lube lightly.
- Run the case into the sizing die to resize and deprime (remove spent primer).

3. **Priming the Case:**
   - Insert a new primer using the press's priming tool or a handheld priming tool.
   - Ensure the primer is seated flush with or slightly below the case base.
4. **Charging with Powder:**
   - Consult your reloading manual for the correct powder type and charge weight.
   - Use a powder measure or digital scale for precise dispensing.
   - Pour powder into the primed case carefully.
5. **Seating the Bullet:**
   - Place a bullet over the case mouth.
   - Run the assembly into the seating die, adjusting depth as needed.
6. **Crimping (Optional):**
   - Apply a light crimp to secure the bullet, especially important for semi-auto firearms.
7. **Inspection:**

   - Inspect each finished round for proper length, primer seating, and overall condition.

Preparing Charges and Measuring Case Length:

- Use calipers to measure case length and overall cartridge length.
- Trim any cases that exceed maximum specifications using a case trimmer.
- Always double-check powder charges to avoid dangerous overpressure.

Safety Measures:

- Always wear safety glasses during reloading.
- Keep powder, primers, and loaded rounds away from heat, sparks, and open flame.
- Never exceed powder charges listed in reputable reloading manuals.
- Label batches clearly with caliber, powder type, charge weight, and bullet weight.

Integration Tip: Link your reloading station with your black powder production and ammo storage areas. By creating a "munitions bench," you centralize all your ammunition maintenance, recycling, and

Materials Needed:

- Spent brass casings (various calibers)
- Bullet puller (kinetic or press-mounted)
- Case trimmer
- Tumbler or ultrasonic cleaner
- Reloading press and dies
- New primers
- Fresh powder or carefully reclaimed powder
- Casting equipment (optional for making new bullets)
- Safety glasses, gloves, dust mask

Safety Equipment:

- Leather gloves for handling sharp, deformed casings
- Heavy apron (optional)
- Fire extinguisher nearby

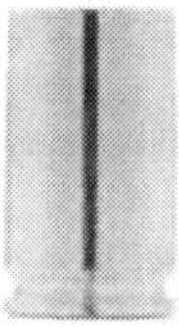

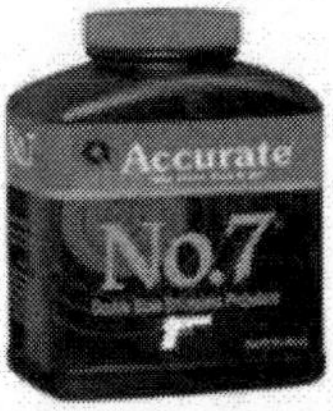

Recycling Procedures:

1. **Collect and Sort Casings:**
    - Gather all spent brass casings after practice sessions or hunting trips.
    - Sort by caliber to streamline reloading later.
    - Discard steel-cased or Berdan-primed cases unless specialized tools are available.
2. **Inspect Casings:**
    - Check for cracks, dents, excessive corrosion, or other signs of structural

weakness.

- Discard any casing that shows damage.

3. **Clean Thoroughly:**
    - Tumble casings with media (corncob, walnut shell) or use an ultrasonic cleaner.
    - Remove carbon build-up and polish brass for smooth reloading.
4. **Trim and Resize:**
    - Use a case trimmer to bring stretched cases back to specification.
    - Resize cases with the appropriate reloading die set.
5. **Bullet Recovery and Reuse:**
    - Use a kinetic bullet puller to safely disassemble misfired or damaged cartridges.
    - Inspect recovered bullets; use if undamaged, remelt if casting your own projectiles.
6. **Powder Recovery (Advanced and Cautious):**
    - Salvage powder from pulled rounds if it's clean, dry, and recently manufactured.
    - Never mix powders from different sources.

    - Label all salvaged powder clearly and use it promptly.
7. **Reloading:**
    - Reprime, repowder, and reseat bullets following standard reloading procedures.

Components You Can Recycle:

- Brass casings (multiple reloads possible)
- Bullets (reuse or melt and recast)
- Powder (in limited, cautious circumstances)
- Packaging materials (ammo boxes, trays, labels)

Components You Should Replace:

- Primers (must always be new)
- Severely corroded, cracked, or damaged brass

Safety Measures:

- Never reuse a primer.
- Never reload brass that shows even minor cracks or bulges.
- Always clearly label recycled ammo to track reload cycles and history.
- Keep meticulous records to avoid confusion or dangerous overpressure

loads.

Integration Tip: Integrate ammo recycling into your larger munitions bench alongside black powder production and cartridge reloading. Pair with metal casting stations to produce new projectiles and fully close the ammunition sustainability loop on your property.

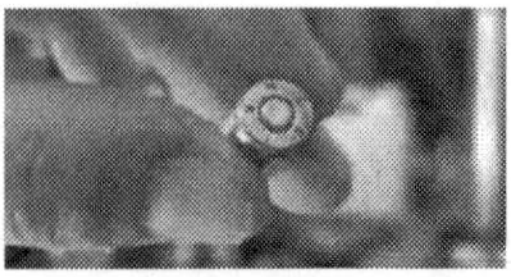

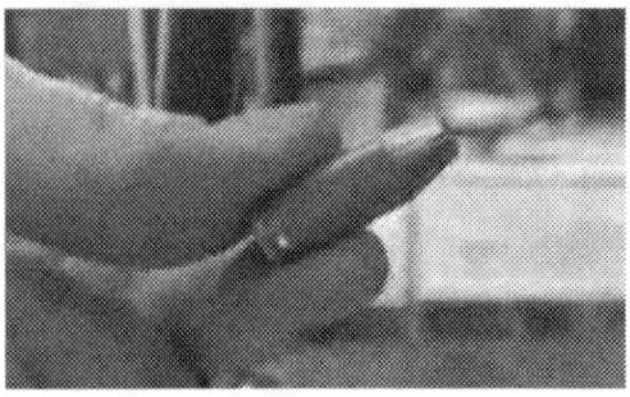

Benefits:

- Greatly extends limited ammunition supplies
- Reduces dependency on external markets
- Turns "waste" into valuable resources
- Supports barter and trade opportunities in

post- collapse scenarios

Sustainability Focus: Use salvaged lead from wheel weights, plumbing scrap, or fishing sinkers to cast new bullets. Trade or barter for used brass casings at ranges or with other preparedness-minded individuals to maintain a constant supply of raw materials.

III. Home & Property Security

A. Intruder Deterrence & Surveillance

Securing your home and property is a vital part of building a fully self-sufficient lifestyle. In an off-grid environment, you cannot depend on public services for protection.
Instead, proactive measures — from strategic lighting to simple yet effective alarms — form your first and often Intruder Deterrence Surveillance best line of defense.

Security should be layered, blending passive deterrents (like lighting and barriers) with active detection (like tripwire alarms and surveillance). By combining multiple approaches, you create a dynamic, resilient security network that protects your water, energy, food supplies, and loved ones.

Each project in this section focuses on low-cost, DIY methods using sustainable, repurposed, or easily accessible materials. Integration into your broader survival systems is emphasized, creating a holistic shield around your property.

Let's begin by shining a light on one of the simplest but most powerful deterrents: motion-activated outdoor lighting.

Chapter 21:

Motion-Activated Outdoor Lighting

Darkness provides cover for potential intruders. Effective outdoor lighting eliminates hiding spots, signals vigilance, and strengthens perimeter security. Motion-activated lighting, in particular, combines energy efficiency with maximum psychological impact: a sudden burst of light startles intruders, alerts you, and gives you precious seconds to respond.

This project guides you through setting up a sustainable, renewable-powered motion lighting system tailored for off- grid properties.

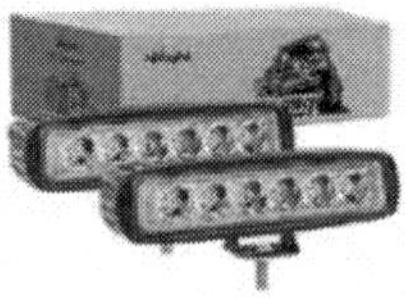

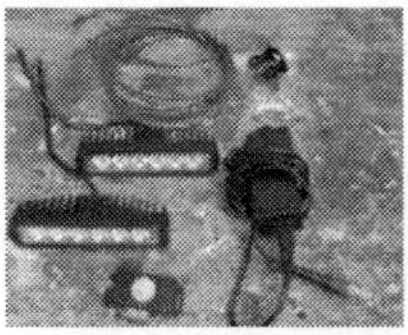
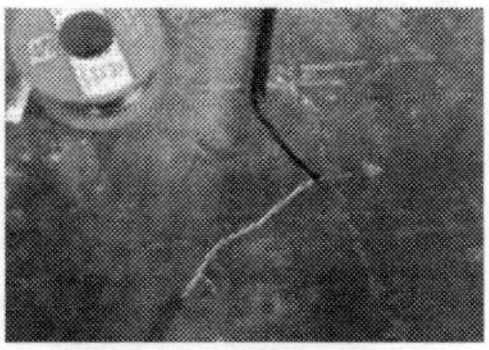
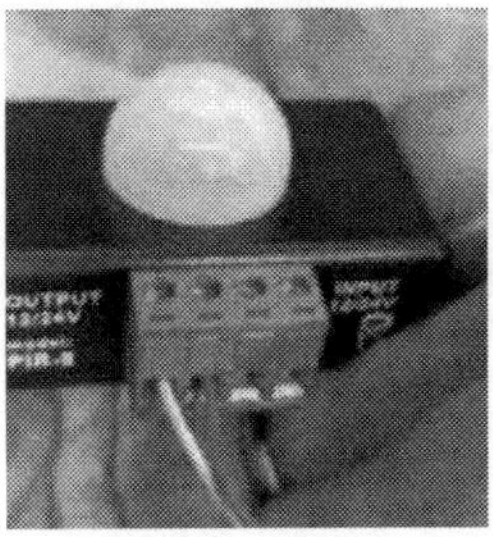
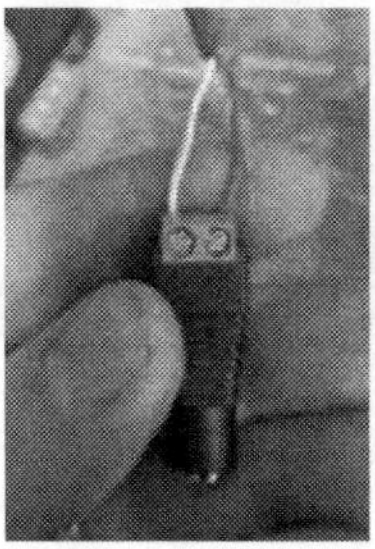
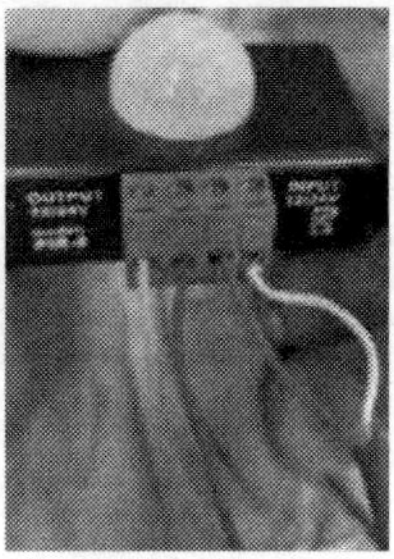

Materials Needed:

- Solar-powered motion-activated floodlights or security lights
- Mounting hardware (screws, brackets)
- Optional: secondary battery packs for extended operation

- Tools: drill, screwdriver, level, ladder, safety harness (for high mounts)

Choosing the Right Lights:

- Prefer solar-powered models with backup batteries for true off-grid operation.
- Look for weatherproof (IP65 or higher

rated) and wide-angle detection (120° to 180°) models.

- Select adjustable units to control detection range, light duration, and brightness.

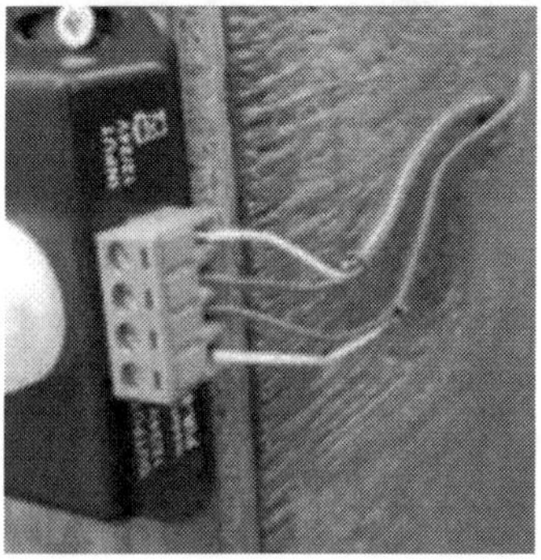

Building Process:

1. **Survey and Plan:**
    - Walk your property at night.
    - Identify vulnerable points: doors, windows, driveways, garden gates, animal pens.
2. **Select Locations:**
    - Mount lights 8–10 feet high for broad coverage.

    - Overlap light fields to avoid dark "blind spots."

3. **Mount the Lights:**
    - Drill and secure mounting brackets into solid surfaces (wood, concrete, brick).
    - Install the light heads securely and angle them downward for maximum ground coverage.
4. **Optimize Detection Zones:**
    - Adjust motion sensors to avoid false triggers from small animals or trees.
    - Test trigger range and re-aim as needed.
5. **Integrate with Other Systems:**
    - Pair lighting with tripwire alarms, camera systems, or automatic recordings to create a layered response network.

Sustainability Focus:

- Use solar models to avoid drawing from your main power system.
- Choose lights with replaceable batteries to extend lifespan.
- Repurpose old mounting poles, fence posts, or

scrap metal for light stands.

Safety Measures:

- Always shut off electrical circuits before mounting hardwired models.
- Use a safety harness and spotter if installing at heights.
- Waterproof all wiring connections to prevent corrosion.

Integration Tip: Pair motion-activated lights with passive barriers like thorny hedgerows or low fences. Light alone deters many threats, but light combined with physical obstacles dramatically increases security without ongoing energy demands.

Benefits:

- Deters intruders without requiring constant power draw
- Extends property visibility and situational awareness
- Enhances safety for nighttime chores or emergencies
- Simple, low-cost installation with scalable coverage

Sustainability Focus: Salvage older solar garden

lights and rewire or combine panels to create larger solar arrays. Extend system life by replacing rechargeable batteries and avoiding disposable units.

Chapter 22:
DIY Razor Wire Installation

While lighting and alarms help deter casual intruders, physical barriers create serious obstacles for determined threats. Razor wire is one of the most effective low-tech security measures available — lightweight, durable, and psychologically intimidating.

Installing DIY razor wire not only strengthens your property's defenses but also integrates seamlessly with your broader perimeter strategies: fencing, motion lighting, tripwire alarms, and surveillance systems.

This project outlines how to install razor wire safely and sustainably, maximizing its effectiveness without drawing unnecessary attention.

Materials Needed:

- Razor wire coils or concertina wire
- Heavy-duty gloves (puncture-resistant)
- Eye protection
- Steel T-posts, wooden fence posts, or repurposed poles
- Heavy-gauge tie wire or fence staples
- Post driver or digging tools
- Wire cutters
- Ladder (if installing at height)
- Warning signs (optional, for visible deterrent effect)

Tools Needed:

- Post driver or sledgehammer
- Pliers
- Wire tensioner (optional for tight installations)
- Level

Building Process:

1. **Plan the Installation:**
    - Identify critical zones to secure: perimeters, choke points, entrances, vulnerable back sections.
    - Decide between ground-level installation (ankle traps) or elevated barriers (fence toppers).

2. **Install Fence Posts:**
    - Drive T-posts or dig holes for wood posts every 6 to 8 feet.
    - Set posts firmly to handle tension and weight.
3. **Prepare the Razor Wire:**
    - Uncoil slowly and cautiously.
    - Wear puncture-resistant gloves and eye protection.
    - If needed, stretch the coil lightly for even spacing.

4. **Mount the Razor Wire:**
   - Secure wire to posts with heavy tie wire or fencing staples.
   - For ground barriers, mount wire 6 to 12 inches above ground to deter crawling intruders.
   - For fence toppers, mount horizontally or at a 45-degree outward angle.
5. **Tension and Secure:**
   - Pull wire snug but avoid over-tensioning (which can snap the wire).
   - Add extra fastening at corners and termination points.

6. **Visibility Considerations:**
   - Install warning signs if laws or location demand disclosure.
   - Hidden barriers may be used in less- trafficked areas for stealth deterrence.

Variations:

- **Single Coil:** Simplest, lightweight deterrent.
- **Double Coil Concertina:** Stronger, more

resilient against cutting or breaching.

- **Triple Strand Fencing:** Three parallel runs of wire for maximum delay effect.

Safety Measures:

- Always work with at least one partner when handling razor wire.
- Wear puncture-resistant gloves and heavy-duty clothing.
- Keep a fully stocked first-aid kit nearby during installation.
- Secure ends of cut wire sections to prevent snagging hazards.

Integration Tip: Combine razor wire with motion-activated lighting and tripwire alarms at key approach points. An intruder startled by lights, hindered by wire, and identified by alarms faces overwhelming odds — maximizing deterrence while giving you precious reaction time.

Benefits:

- Physically delays intruders, buying valuable response time
- Provides psychological deterrent against casual and professional threats

- Long-lasting with minimal maintenance required
- Highly scalable for properties of any size

Sustainability Focus: Whenever possible, source reclaimed razor wire from decommissioned security installations. Reinforce posts with salvaged steel or recycled hardwood beams to avoid purchasing new materials unnecessarily.

Chapter 23:

Building a Trip Wire Alarm System

An early warning can be the difference between safety and disaster. A simple trip wire alarm system acts as your invisible sentry, alerting you to intruders long before they reach critical areas of your property.

Trip wire alarms are low-tech, highly effective, and perfectly suited for off-grid living. They require no electricity, minimal investment, and can be adapted to suit various terrains and threat levels.

This project walks you through building reliable, discreet trip wire alarms using easily sourced, sustainable materials. When layered with your lighting, fencing, and surveillance systems, these alarms create a powerful, proactive defense grid.

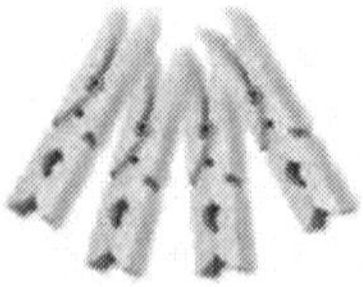

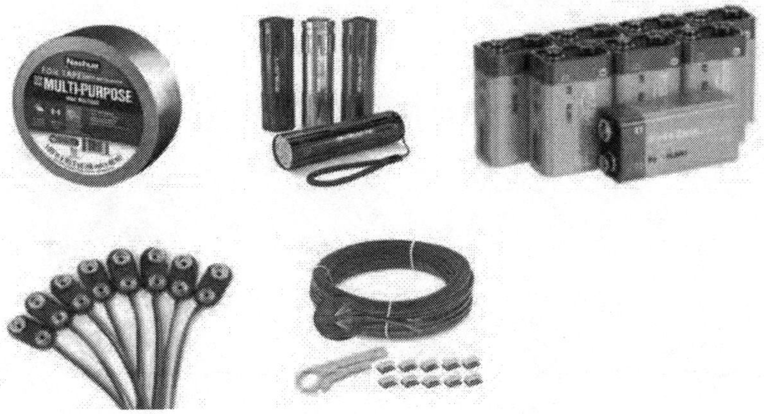

Materials Needed:

- Fishing line, paracord, or fine wire (non-stretch and weather-resistant)
- Clothespins or alligator clips
- Small bells, air horns, or battery-operated buzzers
- AA batteries (for buzzer setups)
- Tent stakes, screws, nails, or eye hooks (for securing lines)
- Zip ties or electrical tape
- Optional: spare tin cans, string, mousetraps, party poppers (for noise-making variations)

Tools Needed:

- Hammer
- Drill (for permanent installations)

- Screwdrivers
- Pliers or wire cutters

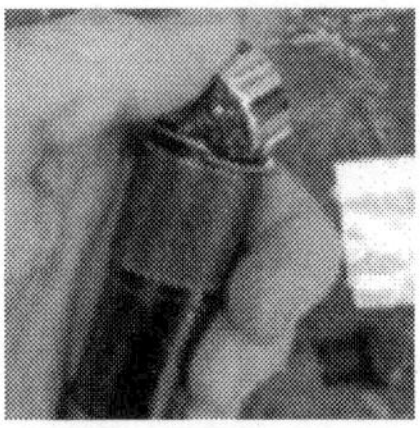

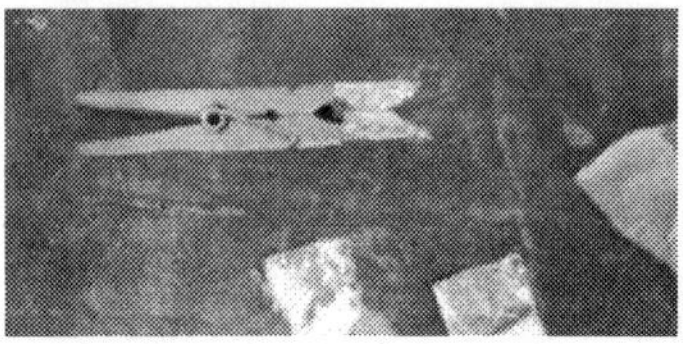
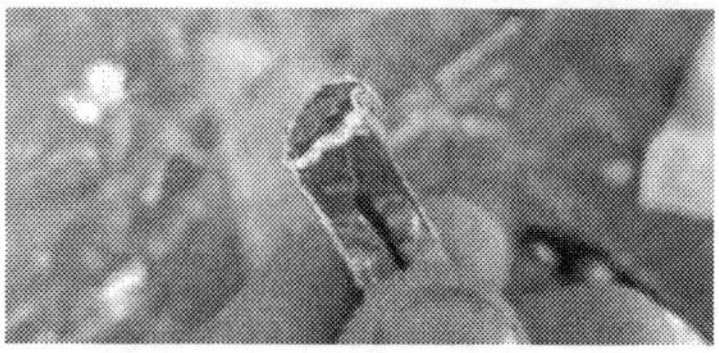
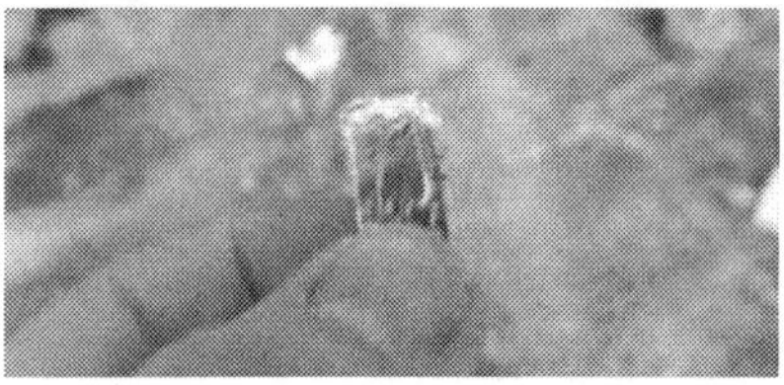
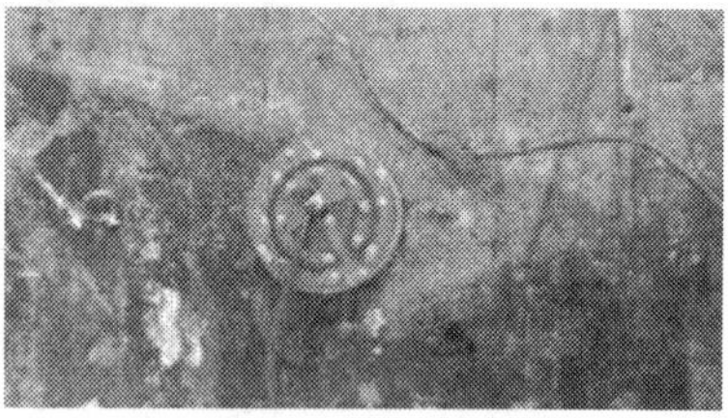

Building Process:

1. **Select the Alarm Type:**
   - **Silent Alarms:** Snap a clothespin-based circuit to activate a buzzer or radio transmitter.
   - **Audible Alarms:** Trigger bells, cans, air horns, or noise makers.

2. **Choose Strategic Locations:**
   - Focus on paths, gates, garden entrances, building perimeters, and hidden approach routes.
   - Conceal trip wires with natural materials (grass, brush) without making them invisible to your family.
3. **Install Anchor Points:**
   - Drive stakes, nails, or eye hooks into the ground, trees, or fences.
   - Place anchor points at ankle to knee height for maximum effectiveness.
4. **Attach the Trip Wire:**
   - Stretch the fishing line or wire between anchor points.
   - Keep slight tension but avoid snapping

tight (which could reveal the wire through movement).

5. **Connect the Alarm Mechanism:**
   - **Clothespin Silent Alarm:**
     - Attach conductive foil strips to each side of a clothespin.
     - Connect to a battery and buzzer circuit.
     - Trip wire pulls the clothespin open, breaking the circuit and triggering the buzzer.
   - **Bell or Can Alarm:**
     - Hang bells or cans loosely.
     - Trip wire tension causes them to rattle loudly.
   - **Party Popper Alarm:**
     - Tie the popper string to the trip wire.
     - Pulling the wire triggers a loud pop.
6. **Test and Adjust:**
   - Trip the wire intentionally.
   - Fine-tune tension and placement for reliability.

Clothes-Pin Silent Alarm Setup:

- Easiest electrical build.
- Use aluminum foil, tape, and lightweight wires.
- Keep circuits weatherproofed with plastic bags or shrink-wrap.

Sound Grenade Alarm (Advanced Variation):

- Attach a pull-pin sound grenade (like a marine distress alarm) to a taut trip wire.
- Extreme noise (120+ dB) deters intruders and alerts property owners over large distances.

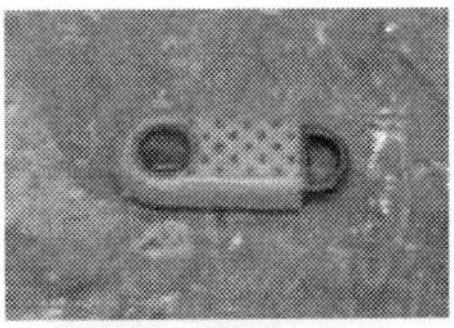
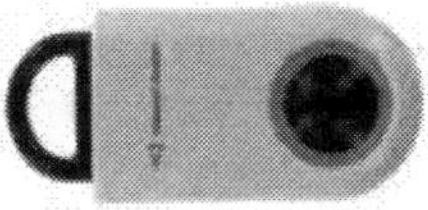

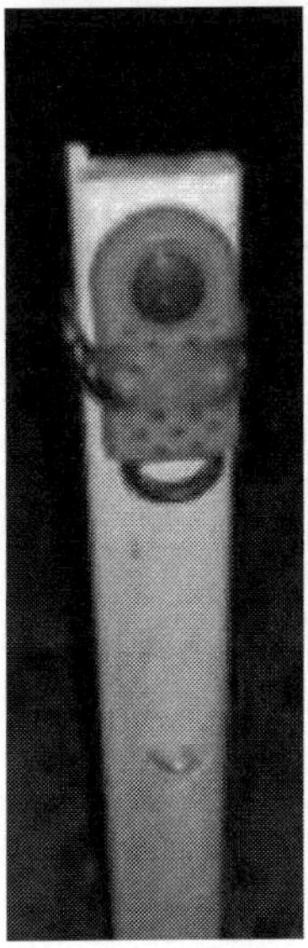

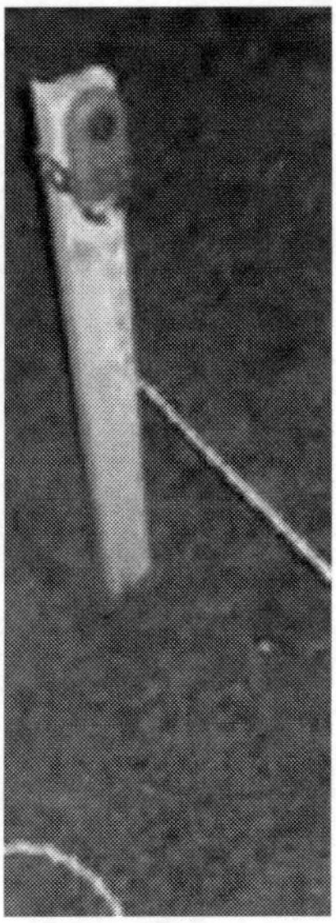

Safety Measures:

- Clearly mark trip wire locations on property maps for family and allies.
- Use visible indicators (small cloth strips) if necessary to avoid accidental triggering by friendly individuals.
- Always disarm alarms before doing maintenance in secured zones.

Integration Tip: Position trip wire alarms to create **layered early warning zones**. Link alarms near motion- activated lights, razor wire barriers, and communication hubs. This synergy ensures maximum detection at minimal cost, buying you critical time to respond or reposition.

Benefits:

- Low-cost, highly effective intrusion detection
- No reliance on external power grids or electronics
- Easy to deploy, reposition, and conceal
- Adaptable to many terrains and property layouts

Sustainability Focus: Repurpose fishing line, old electronics (buzzers, bells), and scrap metal for alarm construction. Use biodegradable or recyclable materials for environmental responsibility when setting temporary perimeter systems.

B. Communication for Emergencies

Chapter 24: CB/FRS/GMRS Communication Systems

In a true off-grid survival scenario, information is as vital as food, water, and energy. Maintaining communication lines between family members, neighbors, and mutual aid networks can mean faster responses to threats, better resource management, and improved morale.

While high-tech options like cell phones or internet-based tools often fail during grid-down events, simple radio systems remain reliable. Citizens Band (CB), Family Radio Service (FRS), and General Mobile Radio Service (GMRS) radios offer flexible, low-cost, independent communication options perfectly suited for off-grid living.

This project outlines how to select, set up, and maintain a resilient radio communication network using CB, FRS, and GMRS technologies.

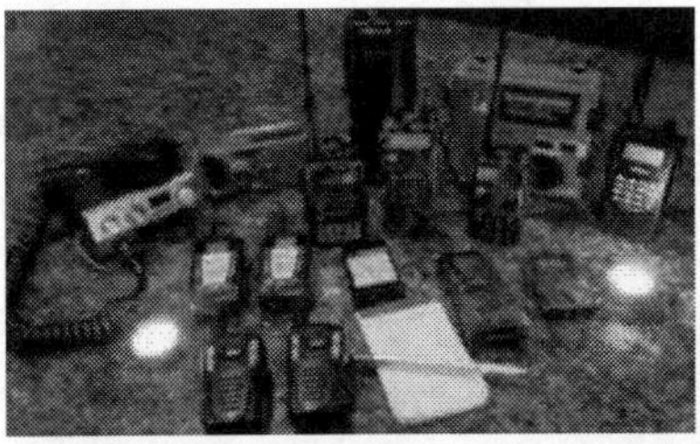

Materials Needed:

- CB radios (base stations, mobile units, or handhelds)
- FRS/GMRS handheld radios
- External antennas (whip or base-mounted)
- Coaxial cables and connectors
- Solar-powered battery chargers or 12V systems
- SWR meter (for CB antenna tuning)
- Programming cables (optional, for GMRS repeaters)
- Waterproof storage cases (for backup radios)

Transmitters, Receivers, and Check-In Procedures:

1. **Choosing the Right Radio Types:**
    - **CB Radios:**
        - No license required.
        - Operates on 40 AM channels around 27 MHz.
        - Great for local communication (1-5 miles normally, up to 20+ miles with upgrades).
    - **FRS Radios:**
        - No license required.
        - Handheld, short-range (up to 2 miles).
        - Excellent for property-wide or neighborhood check-ins.
    - **GMRS Radios:**
        - Requires an inexpensive FCC license.
        - Higher power (up to 50W) and repeater capability.
        - Best for regional communication (up to 20+ miles with repeaters).
2. **Setting Up Base Stations:**
    - Mount CB and GMRS base units inside secure, elevated structures.
    - Install external antennas on

rooftops or towers for maximum range.

  - Connect radios to your off-grid power system (solar battery banks or backup generators).

3. **Programming and Tuning:**
   - Use an SWR meter to properly tune CB antennas and avoid signal loss.
   - Program GMRS radios to match local repeater frequencies if available.
   - Pre-load common emergency channels for fast access.
4. **Establishing Check-In Procedures:**
   - Designate daily check-in times with family or community groups.
   - Use clear, simple call signs.
   - Rotate frequencies periodically to prevent eavesdropping.
5. **Redundancy Planning:**
   - Store backup handheld radios in Faraday cages to protect against EMPs.
   - Keep extra antennas, cables, and batteries on hand.

Safety Measures:

- Maintain radio discipline: no unnecessary chatter during emergencies.
- Encrypt sensitive information through simple code words or prearranged signals.
- Test radios weekly to ensure functionality.

Integration Tip: Link your radio network with tripwire alarms and perimeter lighting. For example, an alarm trip could trigger a call-in protocol across your CB or GMRS network, coordinating fast defensive responses across multiple parts of your property or with neighboring allies.

Benefits:

- Maintains local and regional communication even without cell service
- Supports emergency coordination, resource management, and situational awareness
- Easily expandable with more radios, antennas, or repeaters
- Operates independently of fragile modern infrastructure

Sustainability Focus: Purchase used CB and GMRS equipment from hamfests, estate sales, or surplus auctions. Salvage antennas and coaxial cables where possible to reduce costs and environmental impact. Power all systems with renewable energy setups for true communication independence.

Chapter 25: Preparing for a Blackout

A blackout isn't just a temporary inconvenience when you're living off-grid — it can disrupt critical systems like water pumping, refrigeration, communication, and perimeter defenses. Preparing for blackouts ensures that your off-grid lifestyle remains resilient and operational, even when external or internal systems fail.

This project details simple, effective strategies to maintain functionality during prolonged power losses, focusing on communication readiness, neighborhood coordination, and essential system backups.

Materials Needed:

- Battery-powered radios (AM/FM, NOAA weather band, CB, FRS, GMRS)
- Solar chargers and portable power banks
- Backup lighting (solar lanterns, crank flashlights)
- Extra batteries (rechargeable preferred)
- Emergency power stations (solar generators or DIY battery banks)

- Hand-crank phone chargers (optional)
- Hard-copy emergency contact lists and frequency sheets
- Local maps (paper copies)
- Signal flags, whistles, or mirrors (for non-electronic signaling)

Emergency Frequencies and Neighbor Check-Ins:

1. **Compile Critical Frequencies:**
   - NOAA weather radio channels
   - Local and regional emergency services frequencies
   - Pre-designated CB, FRS, and GMRS channels for family and community groups
2. **Create Hard-Copy Communication Plans:**
   - Write down frequencies, call signs, and procedures.
   - Laminate sheets for durability and weather resistance.
3. **Set Regular Check-In Times:**
   - Daily radio check-ins ensure continued communication even when voice calls aren't possible.

- Rotate who calls first to test everyone's readiness.

4. **Neighborhood Coordination:**
    - Organize "radio watch" shifts during critical events.
    - Establish emergency code words or phrases for signaling immediate danger without revealing sensitive information.
5. **Backup Signaling:**
    - Use whistles, mirrors, colored flags, or light signals as redundant backup communication methods.
    - Plan non-verbal signals (example: three long whistle blasts = immediate help needed).

Power Backup Strategies:

- Charge all radios, lights, and communication devices before major storms or anticipated events.
- Use solar chargers and small portable panels to keep batteries topped up.
- Store hand-crank radios and lighting in waterproof containers for rapid deployment.

Safety Measures:

- Clearly label backup equipment and store it in easily accessible locations.
- Practice blackout drills with family or group members to reinforce procedures.
- Never drain backup batteries below 50% unless absolutely necessary to preserve battery lifespan.

Integration Tip: Integrate blackout preparedness into your larger property defense and resilience plan. Link lighting, communication, and security alarms to shared battery banks or hybrid energy systems (solar, wind, hydro). Ensure that a blackout doesn't mean complete darkness or loss of operational awareness.

Benefits:

- Maintains vital communication and situational awareness during power failures
- Strengthens neighborhood support networks
- Reduces panic, confusion, and disorder during blackouts

- Extends survival system reliability through redundancy

Sustainability Focus: Choose rechargeable battery systems powered by solar or wind sources. Reuse durable containers and repurpose small electronics for backup systems whenever possible. Focus on low-draw, high- efficiency gear to stretch limited energy resources farther during crises.

• Off-Grid Living, Structures, & Utility Projects

Off-Grid Construction & Shelter

Chapter 26: Building a Greenhouse

A greenhouse extends your growing season, protects crops from harsh weather, and creates a controlled environment ideal for off-grid food production. Whether you face short summers, unpredictable storms, or simply want year-round harvests, a greenhouse becomes a cornerstone of your self-sufficient homestead.

This project details how to build a simple, sturdy greenhouse using locally available materials, emphasizing sustainability, modular design, and integration with your broader property systems.

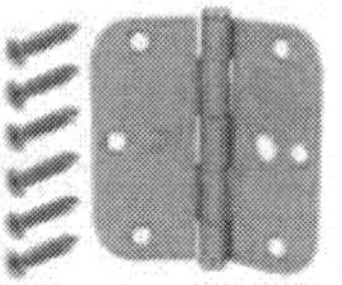
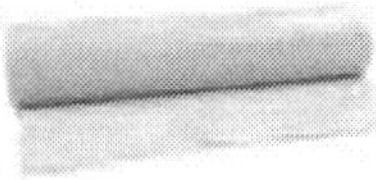

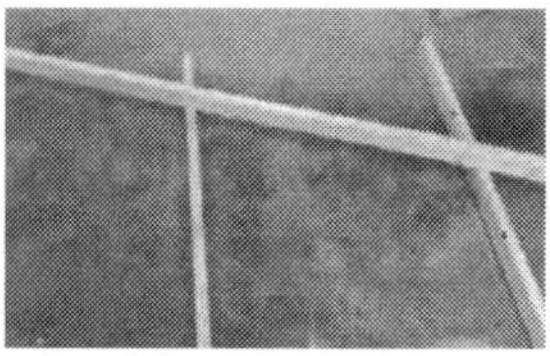

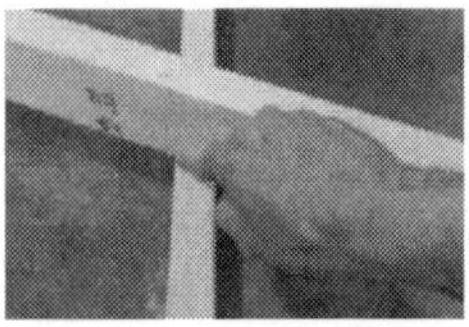

Materials Needed:

- PVC piping (1" or 1.5") or reclaimed metal conduit
- Clear polyethylene plastic sheeting (6 mil UV- resistant)

- Pressure-treated lumber or salvaged wood for the base frame
- Rebar stakes or ground anchors
- Zip ties, clamps, or pipe straps
- Hinges and latches for door assembly
- Basic tools: saw, drill, screwdriver, level, hammer
- Optional: shade cloth, interior shelving, ventilation windows

Design Considerations:

- **Sloped Roof:** Better for shedding rain and snow.
- **Straight Wall Design:** Maximizes internal space.
- **Orientation:** Align east-west for optimal sun exposure.

Building Process:

1. **Site Preparation:**
    - Choose a level, well-drained area with maximum sun exposure.
    - Clear rocks, roots, and debris.
2. **Construct the Base Frame:**
    - Build a rectangular frame from pressure- treated lumber.
    - Anchor frame to the ground using rebar stakes driven through pilot

holes.

3. **Assemble the Hoop Structure:**
   - Bend PVC or metal conduit into arches and attach to the base frame at regular intervals (every 3-4 feet).
   - Use clamps or pipe straps to secure arches firmly.
4. **Install Ridge Pole:**
   - Connect the tops of all hoops with a straight PVC pipe or metal conduit.
   - This ridge pole stabilizes the entire structure.
5. **Attach Plastic Sheeting:**
   - Drape polyethylene sheeting over the frame.
   - Secure with clamps, staples, or furring strips along the base.
   - Leave some slack to accommodate expansion and contraction.
6. **Build and Install Door and Ventilation:**
   - Construct a simple wooden door frame and attach hinges.
   - Cut openings for ventilation windows if desired and cover with mesh or additional plastic.

7. **Seal and Secure:**
    - Patch any small tears immediately with greenhouse repair tape.
    - Ensure all edges are fastened tightly against wind.

Tools and Construction Tips:

- Use a level during frame construction to avoid stress points.
- Pre-bend PVC pipes in warm conditions to reduce cracking.
- For snowy climates, reinforce arches with cross- bracing.

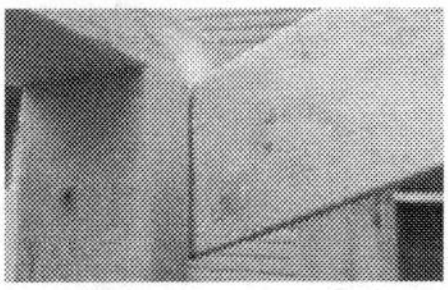

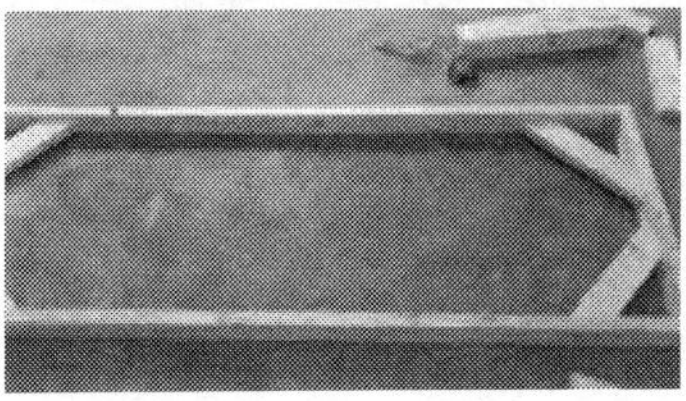

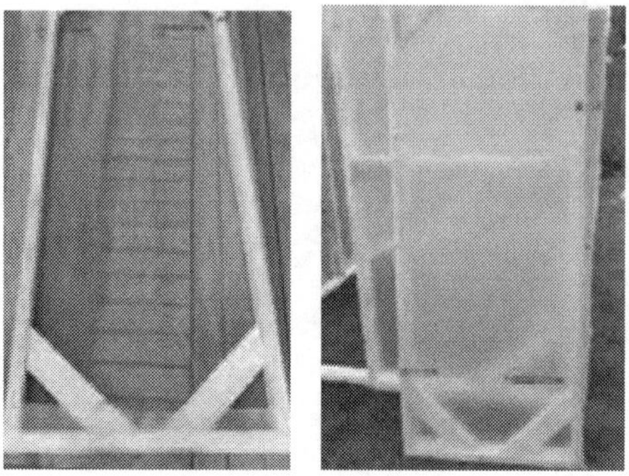

Safety Measures:

- Wear gloves and eye protection when handling plastic sheeting.
- Secure sheeting on calm days to avoid wind hazards.
- Anchor structure thoroughly to prevent tipping in storms.

Integration Tip: Link your greenhouse to rainwater catchment systems for automated irrigation. Combine with your solar power setup to run fans or heating mats if needed. Use adjacent areas for compost piles or worm farms, creating a self-sustaining food production ecosystem.

Benefits:

- Extends growing season year-round
- Protects crops from pests, storms, and

extreme temperatures

- Boosts food security and dietary variety
- Provides seed-starting and medicinal herb cultivation space

Sustainability Focus: Use salvaged lumber, repurposed conduit, and recycled plastic sheeting wherever possible. Create DIY interior shelving from scrap pallets or reclaimed materials to maximize growing space without purchasing new items.

Chapter 27:

How to Build a Hinged Hoop House for a Raised Bed

A hinged hoop house is a flexible, space-saving addition to your off-grid gardening system. Built over a raised bed, it provides protection against frost, pests, and heavy rains while offering easy access to your plants. The hinged design lets you open the structure like a lid, giving you the convenience of an open-air garden when conditions are favorable and instant shelter when needed.

This project maximizes the productivity of raised beds, allowing for earlier planting, extended growing seasons, and stronger, healthier crops — all using affordable, locally sourced materials.

Materials Needed:

- Raised garden bed (pre-existing or DIY-built)
- PVC piping (0.5" to 1")
- Clear polyethylene plastic sheeting (6 mil UV- resistant)

- Hinges (heavy-duty gate hinges or strap hinges)
- Wooden frame (2"x2" or 2"x4" boards)
- Rebar or metal stakes (to anchor PVC hoops)
- Pipe straps, clamps, or zip ties
- Screws, nails, and wood glue
- Basic tools: drill, saw, screwdriver, hammer, level
- Optional: shade cloth, insect netting

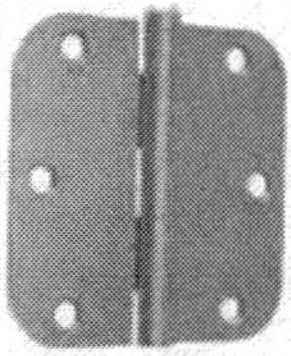

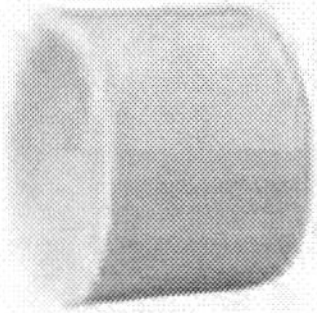

Design Considerations:

- **Bed Size:** Commonly 3'-4' wide x 6'-8' long.
- **Hinge Orientation:** Install on the long

side for larger beds or on the short side for easier lifting.

- **Access:** Ensure the hoop structure is light enough to open with one hand.

Building Process:

1. **Prepare the Raised Bed:**
    - Confirm the bed is level and solidly constructed.
    - Clear planting area of large debris.
2. **Build the Wooden Hoop Frame:**
    - Cut and assemble a simple rectangular wooden frame slightly larger than the bed's opening.
    - Reinforce corners with metal brackets or angled braces.
3. **Install the Hoops:**
    - Hammer rebar stakes into the inside corners of the wooden frame.
    - Slide PVC pipes over stakes to form arching hoops.
    - Space hoops every 2'-3' for strength.
4. **Install the Ridge Pole:**
    - Connect tops of hoops with a lengthwise PVC pipe for added

stability.

5. **Attach Plastic Sheeting:**
    - Drape plastic over the hoop structure.
    - Secure tightly with pipe clamps, zip ties, or by sandwiching between wooden strips and the PVC.
6. **Mount the Hinges:**
    - Attach the hoop frame to the raised bed frame using heavy-duty hinges.
    - Test for smooth opening and closing.
7. **Secure and Seal:**
    - Staple, tack, or glue plastic sheeting at base areas if needed.

    - Add latches or hooks to keep the structure closed during storms.

Tools and Construction Tips:

- Pre-drill holes in wooden frames to prevent splitting.
- Use galvanized or stainless screws for weather resistance.
- Reinforce joints with additional pipe straps if exposed to strong winds.

Safety Measures:

- Wear gloves and eye protection when cutting PVC and plastic sheeting.
- Secure structure properly to avoid accidental tipping.

- Inspect seasonally for tears, loose fittings, or hinge wear.

Integration Tip: Connect your hinged hoop house to a rainwater irrigation system or underground watering pipes. Surround raised beds with pollinator-attracting plants to naturally boost yields and garden health.

Benefits:

- Protects crops from temperature swings and pests
- Allows easy access to plants for watering, pruning, and harvesting
- Extends growing seasons for sensitive crops
- Converts quickly between protected and open-air growing modes

Sustainability Focus: Use reclaimed PVC piping, surplus plastic sheeting, and salvaged lumber to build hoop frames affordably. Patch small holes in coverings with greenhouse tape to extend material life and reduce waste.

Chapter 28: Building a Small Storage Shed from Pallets

Storage is a critical component of any self-sufficient property. Tools, seeds, building materials, and harvested crops all need a dry, secure place to be kept safe from weather and pests. Building a small storage shed from reclaimed pallets provides a low-cost, sustainable, and sturdy solution tailored for off-grid living.

Using pallets minimizes costs, promotes recycling, and allows you to create functional structures without relying on commercial lumber supplies. This project teaches you how to build a basic yet highly adaptable shed perfect for gardens, workshops, or general storage.

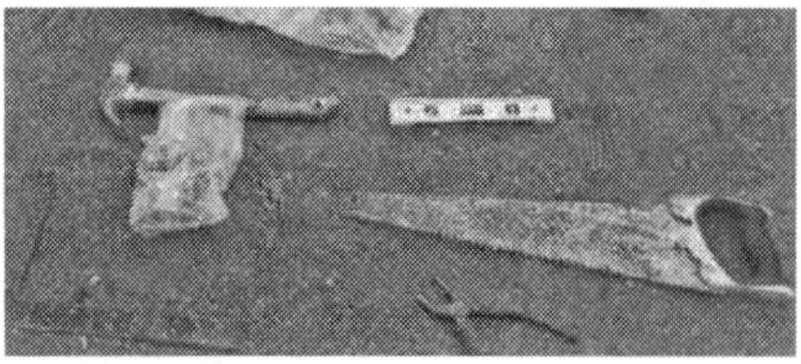

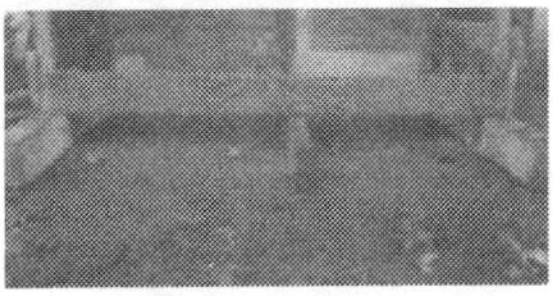
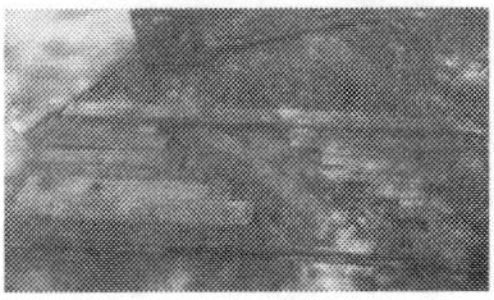

Materials Needed:

- 10-20 wooden pallets (in good condition)
- Pressure-treated lumber or salvaged beams (for framing)
- Corrugated metal sheets, plywood, or reclaimed roofing material
- Screws, nails, bolts

- Metal brackets or braces
- Door hinges and latch
- Clear sealant or exterior paint
- Gravel or concrete pavers (for foundation)
- Basic tools: hammer, drill, saw, crowbar, measuring tape, level
- Optional: plastic sheeting for waterproof lining, insulation material

Design Considerations:

- **Shed Size:** 6'x8' or 8'x10' are manageable for pallet construction.
- **Roof Style:** Sloped shed roof for easy rain runoff.
- **Access:** Wide door opening for easy equipment movement.

Building Process:

1. **Prepare the Site:**
    - Choose a level, elevated spot to prevent flooding.
    - Lay a gravel base or concrete pavers for drainage and stability.

2. **Build the Foundation Frame:**
    - Use treated lumber or strong pallets laid flat.

- Anchor the base securely to the ground.

3. **Erect the Walls:**
   - Stand pallets vertically side-by-side to form walls.
   - Attach pallets together with screws, nails, and metal brackets for extra strength.
   - Reinforce interior corners with scrap lumber or diagonal braces.
4. **Install the Roof:**
   - Frame a simple sloped roof structure.
   - Cover with corrugated metal, plywood, or reclaimed roofing material.
   - Overhang edges to direct rainwater away from the shed base.
5. **Build and Hang the Door:**
   - Construct a door from trimmed pallet boards or repurpose an old solid door.
   - Install heavy-duty hinges and a latch or lock system.
6. **Seal and Weatherproof:**
   - Fill major gaps with scrap wood, caulk, or foam.
   - Apply exterior paint or clear wood

sealant for weather resistance.

Tools and Construction Tips:

- Pre-drill screw holes to prevent splitting older pallet wood.
- Use a pry bar carefully when dismantling pallets for custom fitting.
- Choose pallets stamped "HT" (heat-treated) rather than "MB" (chemically treated) for safety.

Safety Measures:

- Wear gloves and protective eyewear when handling rough pallet wood.
- Check pallets for protruding nails and splinters.
- Stabilize walls and roof sections securely before moving on to next steps.

Integration Tip: Place your storage shed near gardens, greenhouses, or water catchment systems for maximum efficiency. Add gutters to the shed roof to collect additional rainwater for irrigation.

Benefits:

- Provides durable, weatherproof storage space
- Reduces clutter and protects tools, seeds,

and equipment

- Minimizes costs through reuse of free materials
- Customizable and scalable based on property needs

Sustainability Focus: Source pallets from local businesses (hardware stores, manufacturing facilities) where they are often discarded for free. Reclaim roofing, doors, and fasteners from demolition projects to minimize new material purchases and reduce environmental impact.

Chapter 29: DIY Air Conditioner

Staying cool during hot summer months is essential for comfort, health, and productivity on an off-grid property. While traditional air conditioning systems are energy- hungry and grid-dependent, a DIY air conditioner offers a low-power, sustainable solution.

This project teaches you how to create a simple, efficient off-grid cooling system using locally available or repurposed materials. It's ideal for cabins, greenhouses, workshops, and sleeping areas where temperature control can make a real difference.

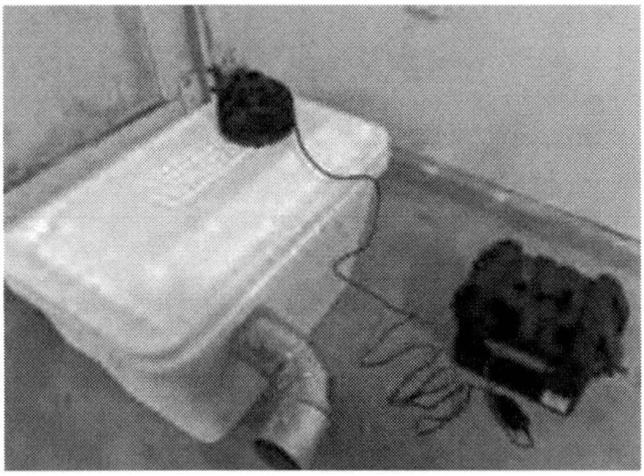

Materials Needed:

- Styrofoam cooler or insulated box
- PVC pipe (elbows and straight pieces)
- Small battery-powered or solar-powered fan
- Ice blocks, frozen water bottles, or ice packs
- Duct tape or adhesive sealant
- Basic tools: knife, drill, hole saw attachment, measuring tape

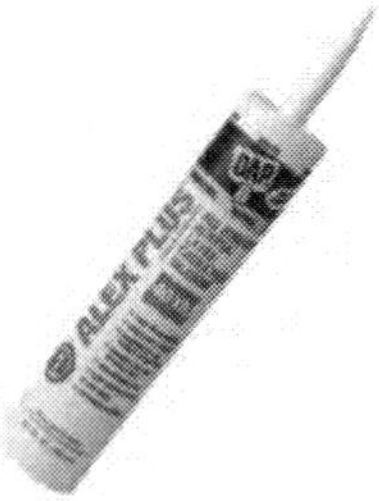

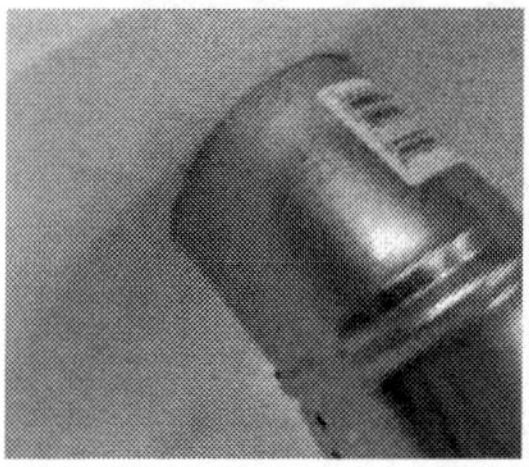
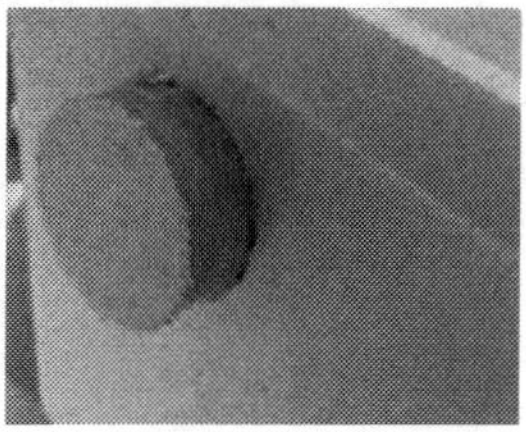

Design Considerations:

- **Cooler Size:** Larger coolers provide longer cooling times.
- **Fan Power:** Low wattage fans extend battery life and efficiency.
- **Ice Source:** Freeze water in reused containers or make large ice blocks in bulk.

Building Process:

1. **Prepare the Cooler:**
    - Choose a well-insulated Styrofoam or heavy-duty cooler.
    - Mark locations for the air intake and outlet.
2. **Drill Air Holes:**
    - Cut two or three holes on the cooler lid — one for the fan, the others for air outlets.
    - Use a hole saw or sharp knife carefully.
3. **Install the Fan:**
    - Mount the fan facing inward over one hole.
    - Secure it with duct tape or adhesive to ensure an airtight seal.
4. **Install the PVC Outlets:**

- Insert short sections of PVC pipe into the outlet holes.
- Angle slightly outward to direct cool air flow.
- Seal edges with tape or caulk.

5. **Fill with Ice:**
   - Place ice blocks or frozen bottles inside the cooler.
   - Arrange for maximum airflow across the ice.
6. **Test and Adjust:**
   - Power on the fan and feel for cold air exiting the pipes.
   - Add or reposition ice as needed to maximize cooling effect.

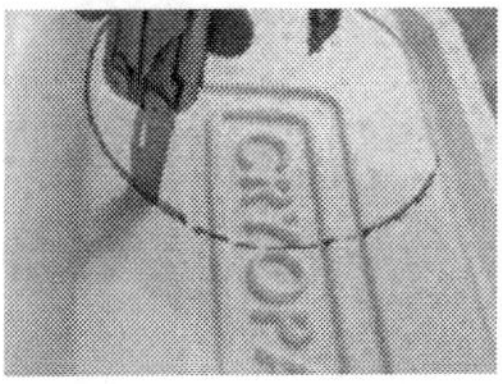

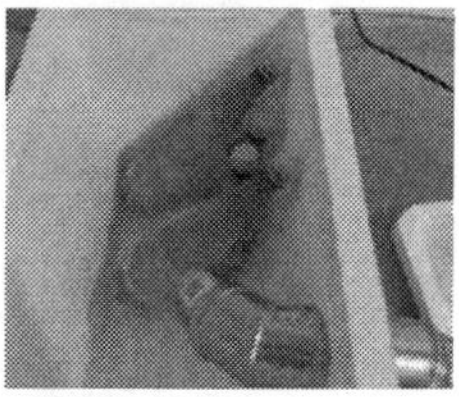

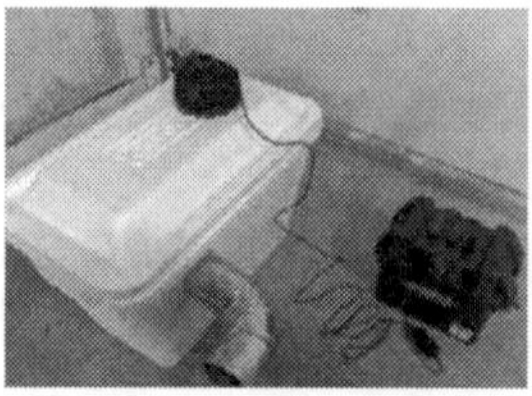

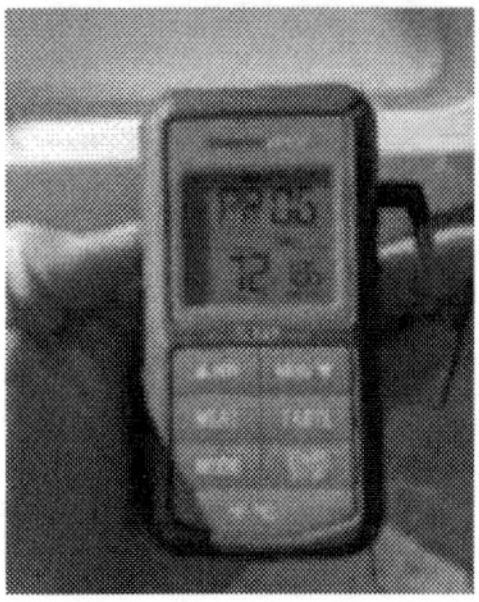
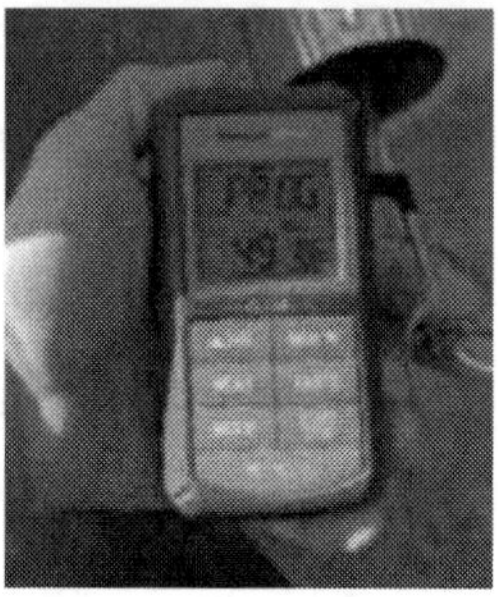

Testing and Variations:

- For longer operation, use larger blocks of ice rather than small cubes.
- Add reflective insulation to the cooler exterior to improve efficiency.
- Mount a solar panel to power the fan independently.

Safety Measures:

- Ensure all wiring is waterproofed if using electrical fans.
- Monitor condensation to prevent puddling around electrical components.
- Use non-toxic containers for making ice if

intended for use around food areas.

Integration Tip: Use your DIY air conditioner in greenhouses to cool delicate seedlings, in workshops to protect temperature-sensitive materials, or in sleeping areas to maintain comfortable rest periods. Integrate it with your solar power grid for a fully renewable cooling solution.

Benefits:

- Provides off-grid cooling without high energy demands
- Increases comfort and productivity during hot months
- Simple, affordable, and scalable
- Easy to repair, upgrade, or adapt to different spaces

Sustainability Focus: Use salvaged coolers, repurposed PVC piping, and old computer or ventilation fans to build your system. Create ice using off-grid freezers powered by your solar, wind, or hydro systems to close the energy loop completely.

Chapter 30:

DIY Solar Dehydrator and Solar Oven

Harnessing the sun's energy to preserve food and cook meals is a cornerstone of true off-grid living. A DIY solar dehydrator and solar oven allow you to dry fruits, vegetables, and meats for long-term storage, and to bake or cook without using gas, wood, or electricity.

Building these devices from simple, local materials provides sustainable, fuel-free alternatives that integrate beautifully with your food production and preservation systems.

Materials Needed for Solar Dehydrator:

- Wooden frame (scrap wood or pallets)
- Clear plastic sheeting or glass panel

- Black metal sheet or painted plywood (for heat absorption)
- Food-grade mesh screens (metal or plastic)
- Hinges and latches (for doors)
- Screws, nails, staples
- Basic tools: drill, saw, hammer, measuring tape, screwdriver
- Optional: small fan for improved air circulation (solar-powered)

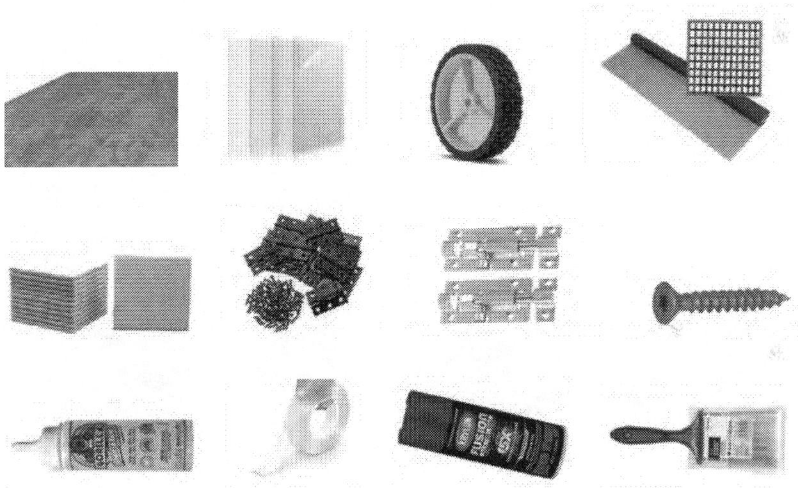

Materials Needed for Solar Oven:

- Cardboard boxes (one large, one slightly smaller)
- Aluminum foil
- Clear glass or plastic sheet
- Black pot or cooking tray
- Insulating material (newspaper, wool, straw)

- Tape, glue, scissors, utility knife
- Wooden stick (to prop up lid)

Building a Solar Dehydrator:

1. **Build the Frame:**
    - Construct a box-like frame tilted at a 30–45 degree angle toward the sun.
    - Install a black metal sheet or black-painted plywood at the bottom to absorb heat.
2. **Install Screens:**
    - Attach food-grade mesh trays inside the frame.

    - Space trays evenly to allow warm air circulation.
3. **Cover and Vent:**
    - Install clear plastic sheeting or a glass panel on the front.

- Add vents at the bottom and top for airflow (covered with mesh to keep insects out).

4. **Test and Adjust:**
    - Place sliced fruits or vegetables on trays.
    - Monitor airflow and temperature (ideal drying range: 120°F–140°F).

Building a Solar Oven:

1. **Prepare the Boxes:**
   - Line the inner box with black construction paper or black paint.
   - Line the inside of the outer box with insulating material.
2. **Create Reflective Surfaces:**
   - Cover inner flaps with aluminum foil (shiny side out).
3. **Assemble the Oven:**
   - Place the smaller box inside the larger one, insulating all gaps.

   - Cover the opening with a clear glass or plastic sheet to create a greenhouse effect.
4. **Cooking with the Oven:**
   - Preheat by placing the oven in full sun.
   - Use a black pot or tray to absorb heat efficiently.
   - Use a stick to prop open the top flap to reflect sunlight inside.

Testing and Variations:

- Adjust air vents in dehydrators for faster or slower drying.
- Improve solar ovens with multiple reflective panels for higher temperatures.
- Add a solar-powered fan to dehydrators for better airflow.

Safety Measures:

- Monitor temperatures closely to prevent over- drying or burning food.
- Use food-safe materials throughout construction.
- Secure solar devices firmly against strong winds.

Integration Tip: Place your solar dehydrator near your greenhouse or garden for easy transfer of fresh produce. Use your solar oven for daily cooking, reducing fuel consumption and complementing your energy-independent kitchen.

Benefits:

- Preserves fruits, vegetables, herbs, and meats without electricity
- Provides renewable, fuel-free cooking options
- Reduces dependency on propane, gas, or wood
- Increases overall food security and energy efficiency

Sustainability Focus: Build dehydrators and ovens from reclaimed wood, salvaged glass, and recycled insulation materials. Design modular frames that can be repaired easily with common parts, extending system lifespan and reducing waste.

Chapter 31:
DIY Water Heater

Having access to hot water off-grid is a major comfort and a vital component of hygiene, cooking, and cleaning. A DIY water heater provides a reliable, energy-efficient solution without the need for expensive propane or electric systems. By integrating renewable energy sources like solar heat or wood stoves, you can maintain steady hot water supplies even in remote conditions.

This project walks you through creating a simple, durable water heating system designed for off-grid living.

Materials Needed:

- Black-painted metal or PVC piping (for solar collector)
- Old water heater tank or food-grade steel barrel
- Copper coil tubing (for faster heating)
- Insulated box or solar water heater enclosure
- Basic plumbing fittings and valves
- Hose clamps, pipe straps, Teflon tape

- Pressure relief valve (safety requirement)
- Wood stove (for backup heating)
- Basic tools: drill, saw, pipe cutter, wrenches, screwdriver

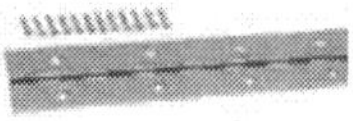
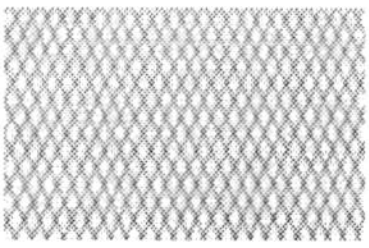

Design Considerations:

- **Solar Heater:** Ideal for sunny climates; completely renewable.
- **Wood Stove Integration:** Essential for cloudy or cold regions.
- **Tank Size:** 30–55 gallons depending on household needs.

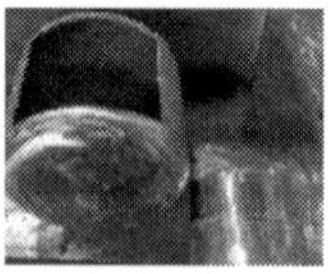
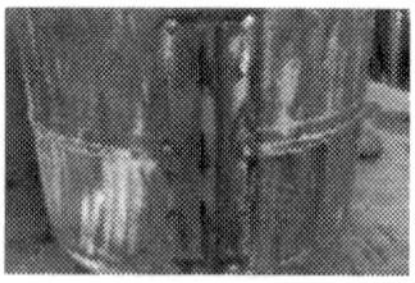

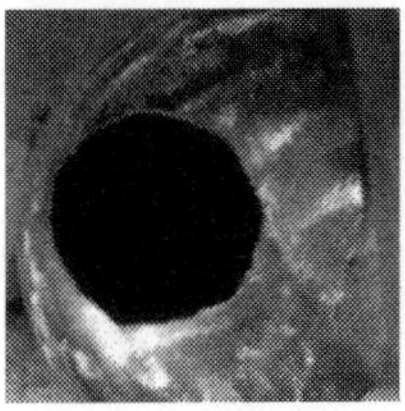
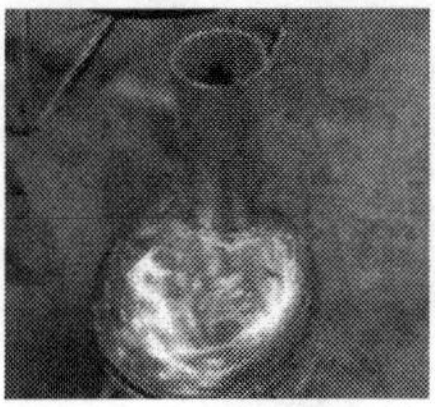

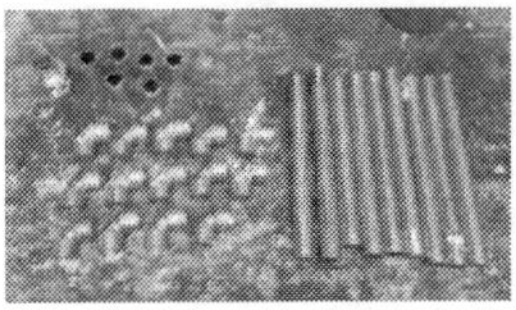

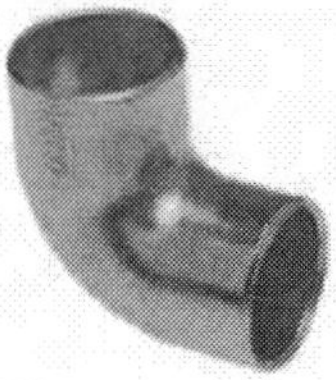

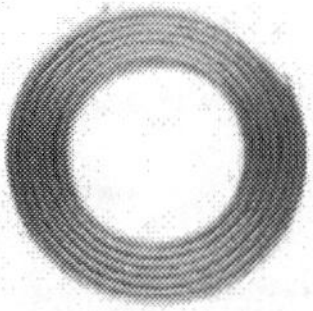

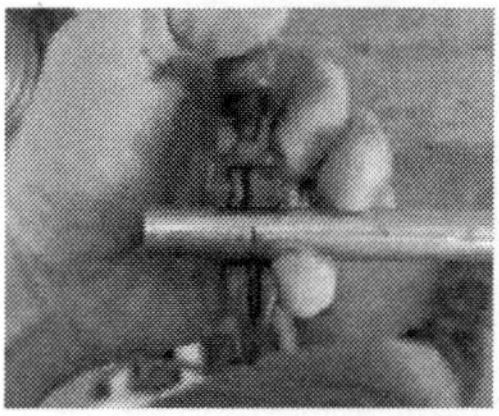
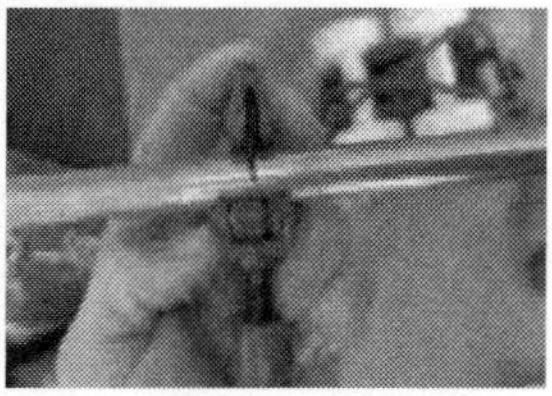

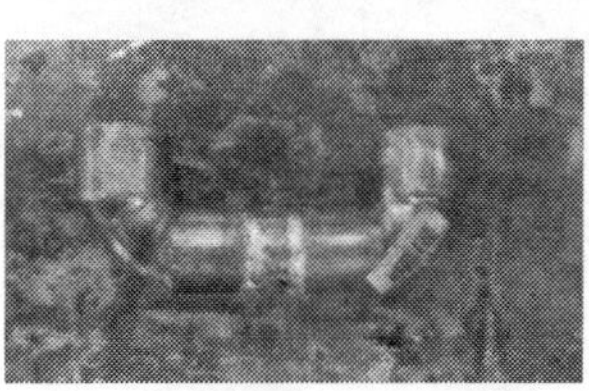

Building Process:

1. **Prepare the Solar Collector:**
    - Paint metal or PVC pipes black for maximum heat absorption.
    - Arrange pipes in a serpentine pattern inside an insulated box.
    - Cover with clear glass or durable plastic sheet to create a greenhouse effect.
2. **Modify or Prepare the Tank:**
    - Clean and sanitize old water heater tanks or barrels.

- Install fittings for cold water input and hot water output.
- Attach a pressure relief valve at the top.

3. **Connect Collector to Tank:**
    - Install copper coil or black piping to circulate between solar collector and storage tank.
    - Use gravity feed (thermosiphon) if possible to avoid pumps.
4. **Install Backup Wood Stove Coil (Optional):**
    - Wrap copper coil around the stovepipe or inside a water jacket.
    - Circulate water through stove coil during colder months.
5. **Secure and Test System:**
    - Mount pipes and tank securely to prevent movement.
    - Fill the system with water and test for leaks.
    - Monitor temperatures and adjust placement for optimal sunlight exposure.

Wood Stove Integration Steps:

- Ensure coils do not restrict stovepipe airflow.
- Use high-temperature rated piping and fittings.
- Always include a pressure relief valve to prevent dangerous steam buildup.

Testing and Variations:

- Adjust tilt angle of solar collector for seasonal sun changes.
- Insulate storage tanks heavily for better nighttime heat retention.
- Use multiple collectors for higher volume systems.

Safety Measures:

- Install a pressure relief valve to prevent tank explosion.
- Insulate hot water lines to prevent burns.
- Monitor temperatures regularly to avoid overheating.

Integration Tip: Connect your water heater output to your off-grid shower system, kitchen, and greenhouse irrigation systems. Use excess heated water for cleaning tools, washing produce, or even running thermal mass heating loops in greenhouses during cold seasons.

Benefits:

- Provides reliable, renewable hot water year-round
- Reduces reliance on propane, electricity, or diesel
- Supports hygiene, food preparation, and comfort
- Integrates easily with other water and heating systems

Sustainability Focus: Use salvaged piping, old water heater tanks, and reclaimed building materials wherever possible. Designing systems without pumps and relying on gravity flow minimizes mechanical complexity and reduces maintenance needs.

Chapter 32: Indoor Greenhouse & Shelf Construction

An indoor greenhouse allows you to grow herbs, vegetables, and seedlings year-round, regardless of the outdoor weather. By constructing simple, modular greenhouse shelves, you can maximize space, improve food security, and create a steady supply of fresh produce right inside your off-grid home.

This project teaches you how to build a compact, efficient indoor greenhouse system using affordable materials and sustainable practices.

Materials Needed:

- Sturdy shelving unit (metal, wood, or PVC)
- Clear plastic sheeting, old window panes, or repurposed glass doors
- Grow lights (LED preferred for low energy use)
- Light timer
- Reflective material (mylar sheets, aluminum foil, or white paint)

- Trays or shallow containers for planting
- Basic tools: drill, screws, measuring tape, utility knife, staple gun

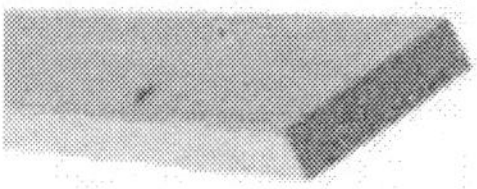

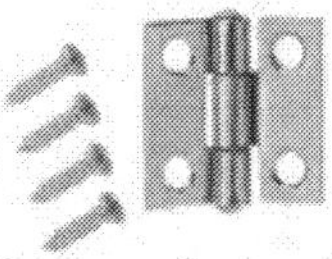

Design Considerations:

- **Size:** Adjust to available indoor space — 2'-4' wide shelves are most manageable.
- **Lighting:** Supplement natural light with grow lights for consistent growth.

- **Temperature and Humidity:** Monitor and adjust as needed for plant health.

Building Process:

1. **Prepare the Shelving Unit:**
    - Choose a shelf unit sturdy enough to hold soil, water, and plants.
    - Clean and reinforce shelves if necessary.
2. **Create the Greenhouse Enclosure:**
    - Cover the back, sides, and top of the shelf with clear plastic sheeting or repurposed glass panels.
    - Staple, tape, or secure the material tightly to trap warmth and humidity.

3. **Install Lighting:**
    - Mount LED grow lights above each shelf or every two shelves.
    - Wire lights to a timer to simulate natural daylight cycles (12-16 hours of light per day).
4. **Improve Light Efficiency:**
    - Line interior surfaces with reflective material to maximize light distribution.
5. **Set Up Plant Trays:**

- Place waterproof trays or containers on each shelf to hold plants.
- Use shallow trays for seedlings and deeper containers for larger plants.

6. **Ventilation (Optional):**
    - Cut small vents into plastic sheeting if excessive humidity becomes an issue.
    - Add a small solar-powered fan for airflow if needed.

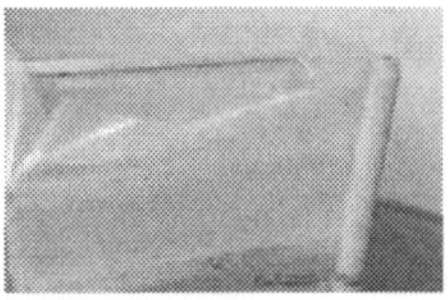

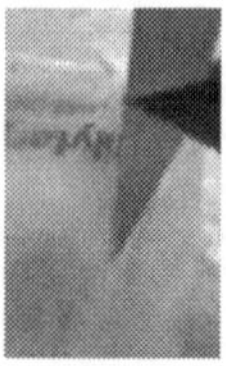
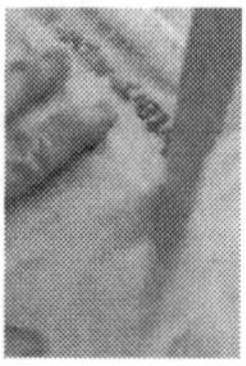

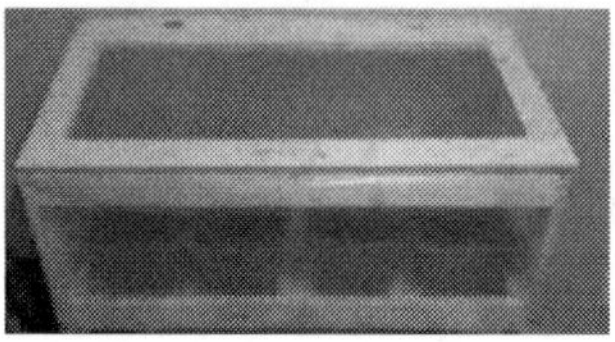

Tools and Construction Tips:

- Pre-plan electrical wiring paths for safety.
- Ensure shelves are level to prevent water pooling.
- Use waterproof power strips for lighting connections.

Safety Measures:

- Keep electrical components away from water sources.
- Secure heavy lights and materials firmly to prevent falls.
- Monitor temperature and humidity regularly to prevent mold or plant stress.

Integration Tip: Combine your indoor greenhouse with your DIY air conditioning and solar power systems to create a fully renewable indoor growing operation. Use your indoor garden to grow culinary herbs, leafy greens, or starter plants for larger outdoor gardens and hoop houses.

Benefits:

- Year-round food production regardless of climate
- Boosts household nutrition and food security
- Increases seed-starting and transplant success rates
- Enhances indoor air quality and natural aesthetics

Sustainability Focus: Repurpose old shelving units, window frames, or closet doors to build your indoor greenhouse affordably. Use LED lighting powered by solar-charged battery banks to minimize energy use and reduce environmental impact.

Chapter 33:

Off-Grid Shower and Stone Grill

Maintaining hygiene and the ability to prepare food outdoors are essential elements of sustainable off-grid living. Building an off-grid shower and stone grill ensures you can stay clean, comfortable, and well-fed without relying on complex infrastructure or energy-intensive systems.

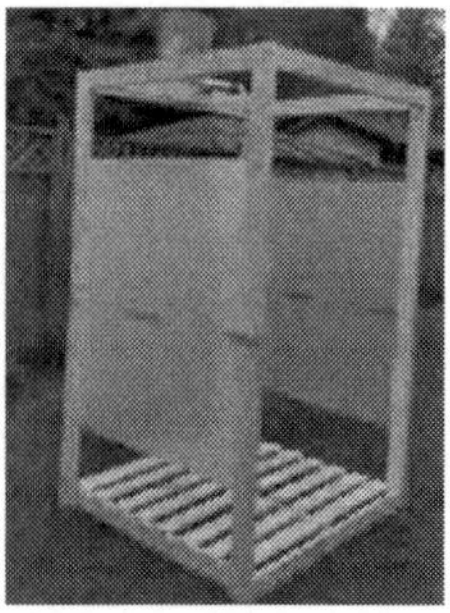

This project guides you through constructing a simple, durable outdoor shower and a robust stone grill using natural and salvaged materials.

Materials Needed for Off-Grid Shower:

- 5-gallon bucket, rain barrel, or recycled drum
- Showerhead or perforated watering can rose

- PVC piping or flexible hose
- Wooden posts or sturdy tree trunks
- Rope, hooks, or pulleys (optional for hanging system)
- Gravel or flat stones (for shower floor drainage)
- Privacy screen materials (tarp, wood slats, bamboo, fabric)

Materials Needed for Stone Grill:

- Large, flat stones (for grill surface)
- Smaller stones for the base
- Metal grill grate (repurposed oven rack, old BBQ grate, or handmade rebar grid)
- Fireproof mortar or clay (optional for added stability)
- Basic tools: shovel, level, hammer, trowel

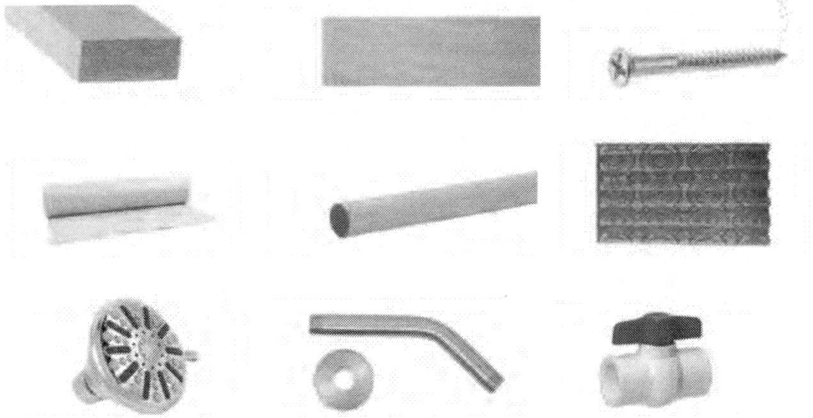

Building the Off-Grid Shower:

1. **Site Selection:**
    - Choose a sunny, well-drained location near your water supply.
    - Ensure the area is private or easily screened.
2. **Prepare the Shower Floor:**
    - Dig a shallow pit and fill with gravel for drainage.
    - Lay flat stones over the gravel for a comfortable standing area.
3. **Construct the Frame:**
    - Install vertical posts to support the water container.
    - Use ropes or hooks to suspend the water source above head height.
4. **Install the Water System:**

    - Attach a showerhead or perforated can rose to a hose or PVC pipe.
    - Connect the hose to the water container.
    - Add a simple valve or spigot to control water flow.

5. **Build the Privacy Screen:**
    - Surround the shower with tarp, fabric, wooden panels, or living bamboo walls.

  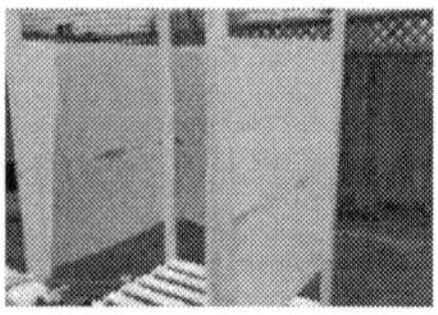

Building the Stone Grill:

1. **Site Preparation:**
   - Clear a level area away from flammable vegetation.
   - Dig a shallow fire pit if desired.
2. **Lay the Foundation:**
   - Stack smaller stones in a U-shape or circular wall.
   - Use fireproof mortar or packed clay for stability if necessary.
3. **Create the Cooking Surface:**
   - Place the grill grate across the top of the stone walls.
   - Ensure stable support at all corners.
4. **Optional Smoke Hood:**

    - Build a simple overhead frame and drape with a tarp or metal sheet to direct smoke away from cooking areas.

Testing and Variations:

- Adjust shower height or flow for different family members.
- Customize grill size based on cooking needs.
- Add solar bag heaters to the shower system for warm water.

Safety Measures:

- Ensure stones are stable and cannot shift during use.
- Clear area around the grill of dry leaves and debris.
- Test all water systems for leaks and secure fittings properly.

Integration Tip: Position your off-grid shower near gardens to allow graywater recycling into irrigation systems. Build the stone grill near your outdoor kitchen,

smokehouse, or solar oven station to create a fully functional off-grid cooking zone.

Benefits:

- Provides hygiene and cooking facilities independent of utilities
- Utilizes local, sustainable materials
- Reduces reliance on propane, electricity, or commercial fuels
- Enhances comfort and self-sufficiency

Sustainability Focus: Source stones from your own property or nearby natural areas. Use salvaged containers and scrap plumbing parts for the shower system. Design both projects for longevity and easy maintenance to reduce future material needs.

Chapter 34: Off-Grid Washing Machines

Laundry can quickly become a major challenge when living off-grid, especially without a reliable energy source.

Handwashing is labor-intensive, but simple mechanical solutions can dramatically reduce effort and water use. A DIY off-grid washing machine brings modern convenience to remote living, helping you maintain hygiene efficiently.

This project outlines two proven designs: a plunger-style washer and a bicycle-powered washer — both easy to build, repair, and integrate into your sustainable home systems.

Materials Needed for Plunger Washing Machine:

- 5-gallon bucket with lid (food-grade preferred)
- New clean toilet plunger or purpose-made mobile washer plunger
- Drill and hole saw bit
- Basic hand tools: drill, knife, screwdriver
- Optional: second bucket for rinse cycle

Materials Needed for Bicycle-Powered Washer:

- Old bicycle (stationary stand or repurposed frame)
- 5-gallon bucket or small barrel with watertight lid
- Pulleys, belts, or direct chain drive
- Reinforced platform or frame
- Drill, saw, wrenches, pliers

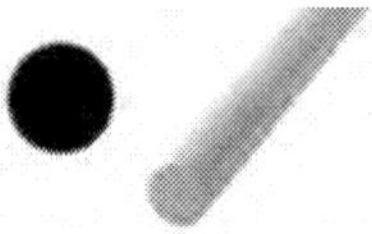
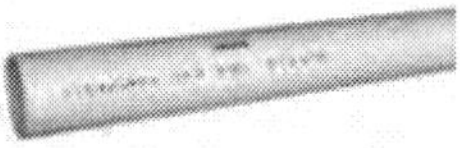
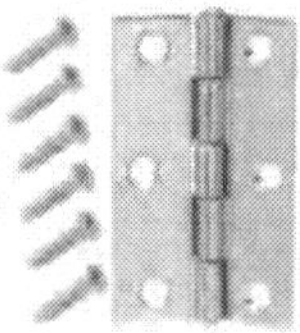
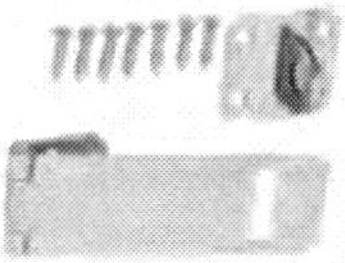

Building the Plunger Washing Machine:

1. **Prepare the Bucket:**
   - Drill a hole in the center of the bucket lid just wide enough for the plunger handle to pass through.
2. **Modify the Plunger:**
   - Drill a few small holes in the rubber plunger head to allow better water and soap circulation.
3. **Assemble the Washer:**
   - Place clothes, soap, and water in the bucket.
   - Insert the plunger through the lid.
   - Secure the lid tightly.

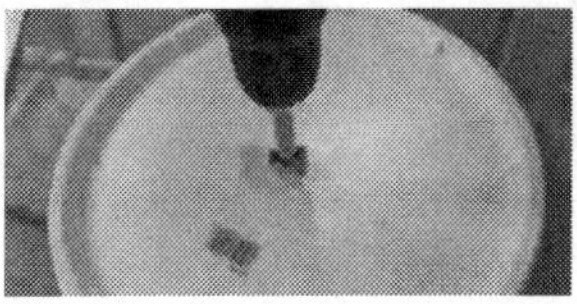
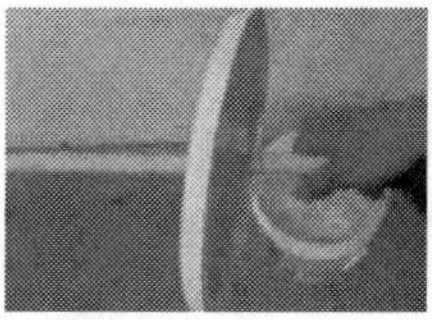

4. **Operation:**
   - Move the plunger up and down vigorously for several minutes.
   - Rinse with clean water and repeat plunging if needed.

Building the Bicycle-Powered Washing Machine:

1. **Prepare the Bicycle Frame:**
   - Remove the rear wheel.
   - Attach a pulley or sprocket to the pedal crank.
2. **Set Up the Bucket Drum:**
   - Secure a 5-gallon bucket or barrel horizontally.
   - Modify it to spin freely using a direct chain drive or belt system from the bicycle pedals.
3. **Connect and Test:**
   - Pedal to rotate the drum, agitating the clothes inside.
   - Adjust tension and rotation speed for optimal washing.
4. **Finishing Touches:**
   - Add internal baffles inside the drum to improve agitation (optional).

Testing and Variations:

- Use hot water heated by your DIY solar water heater for better cleaning power.

- Add a simple drain tap at the bucket base for easier water changes.

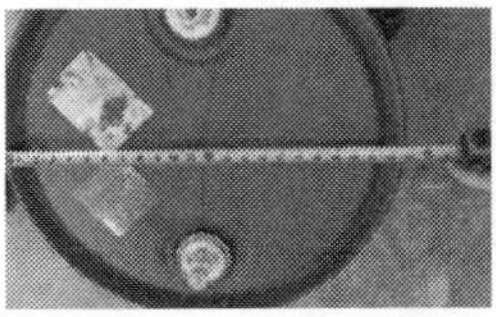

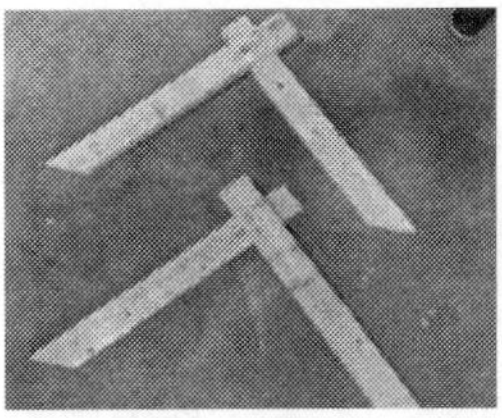

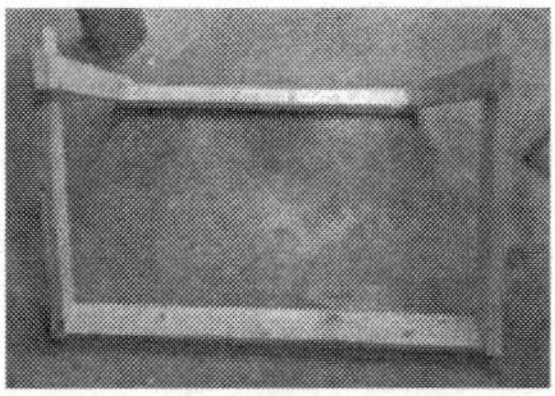

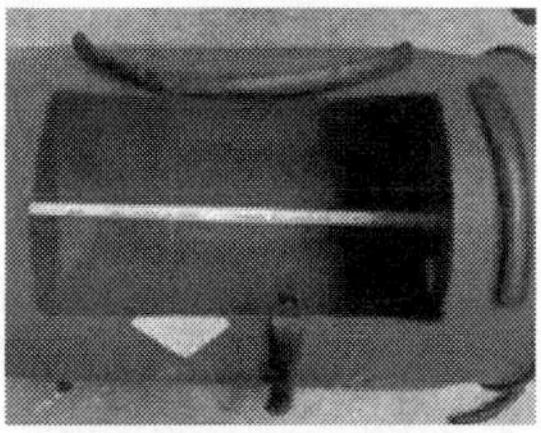
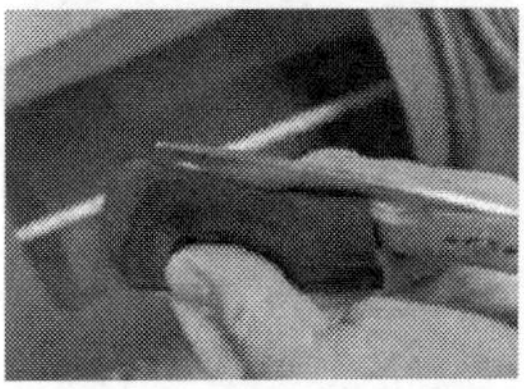

Safety Measures:

- Ensure buckets and lids are free of cracks and secure before operation.
- Anchor bicycle stands or frames solidly to prevent tipping during use.
- Avoid overloading buckets, which can strain equipment and reduce cleaning effectiveness.

Integration Tip: Locate your washing station near your rainwater catchment system and outdoor graywater recycling zone. Use the runoff from rinsing cycles to irrigate trees or non-edible plants, turning laundry into another sustainable water cycle input.

Benefits:

- Drastically reduces water and energy use compared to electric washers
- Empowers self-sufficiency with minimal maintenance
- Supports hygiene and clothing longevity off-grid
- Encourages family or group participation with simple manual operation

Sustainability Focus: Build using salvaged bicycles, discarded food-grade buckets, and reused plumbing parts. Design for easy part replacement and simple hand repairs to extend life and avoid dependence on hard-to-find equipment.

B. Fuel, Bio-fuel, and Waste Utilities

Fuel independence and waste management are pillars of a truly resilient off-grid lifestyle. As traditional energy sources grow more expensive and unpredictable, creating your own renewable fuels and managing waste responsibly becomes essential for maintaining comfort, mobility, and sustainability.

In this section, we explore practical projects for:

- Storing traditional fuels like gasoline and diesel safely and sustainably.
- Producing bio-fuel alternatives from leaves, manure, and other organic materials.
- Turning everyday waste into valuable resources that close the self-sufficiency loop.

Each project is designed not only to solve immediate challenges but also to integrate into your larger property systems, supporting food production, energy needs, and environmental stewardship.

Mastering fuel, bio-fuel, and waste utilities minimizes

external dependence, enhances your property's regenerative capabilities, and ensures long-term resilience no matter what challenges arise.

Let's move forward to create truly independent energy and waste solutions for your off-grid homestead.

Chapter 35: Building a Fuel Storage System

Reliable access to fuel — whether for vehicles, generators, heating, or essential machinery — is crucial for long-term off-grid survival. Establishing a safe, sustainable fuel storage system ensures you have energy reserves during emergencies, seasonal shortages, or supply disruptions.

This project covers best practices for selecting storage containers, safe placement strategies, fuel preservation methods, and integrating your fuel system into your overall off-grid infrastructure.

Materials Needed:

- Approved fuel storage containers (gas cans, metal drums, fuel tanks)
- Fuel stabilizers (for gasoline or diesel)
- Heavy-duty hose, pump, or siphon kit

- Supports resilience in transportation, heating, and power generation

Sustainability Focus: Use repurposed fuel tanks from farms, construction sites, or industrial sources (after thorough cleaning). Salvage pallets or reclaimed bricks for off-ground platforms. Manage inventory meticulously to minimize waste and contamination.

Chapter 36:

How to Make Bio-fuel from Leaves and Manure

In a world where every resource counts, even waste becomes an opportunity. Bio-fuel production from natural materials like leaves and manure offers a renewable, low- cost energy source that fits perfectly into an off-grid lifestyle. Whether used for heating, cooking, or powering small engines, homemade bio-fuel transforms "waste" into a powerful tool for energy independence.

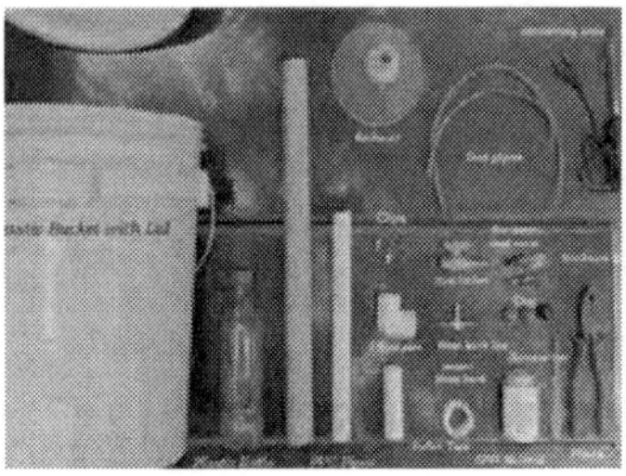

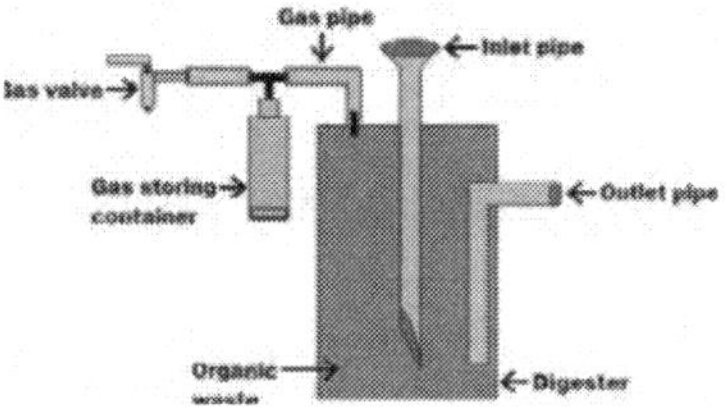

This project teaches you how to safely and effectively produce basic bio-fuel with simple, sustainable methods.

Materials Needed:

- Green leaves (fresh or recently fallen)
- Animal manure (preferably from herbivores like cows, goats, or rabbits)
- Water
- Airtight barrel, drum, or large bucket with lid
- Gas collection hose and storage bladder (optional for biogas)

- Mixing tools (shovel, stick)
- Basic tools: drill, rubber gaskets, hose clamps
- Optional: compost thermometer

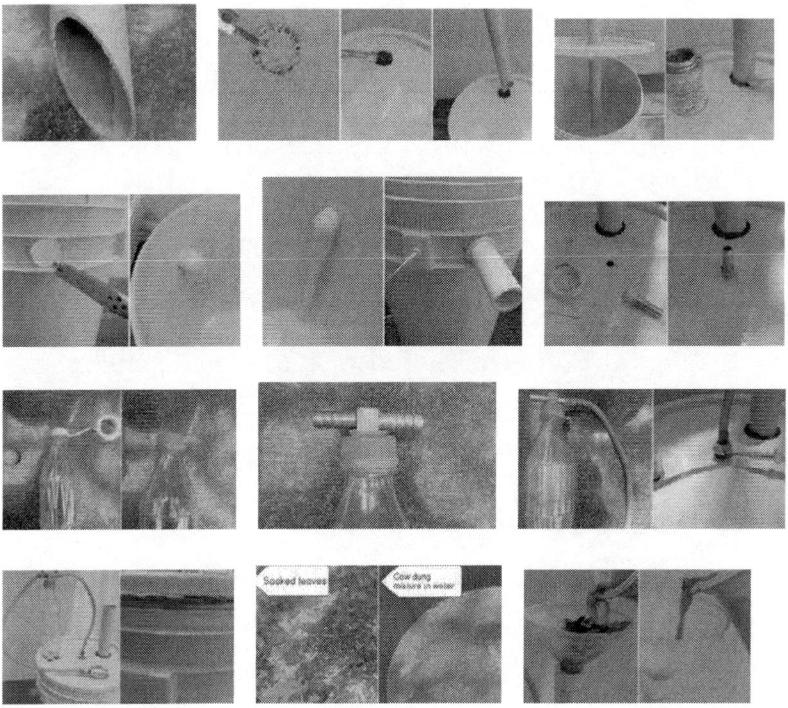

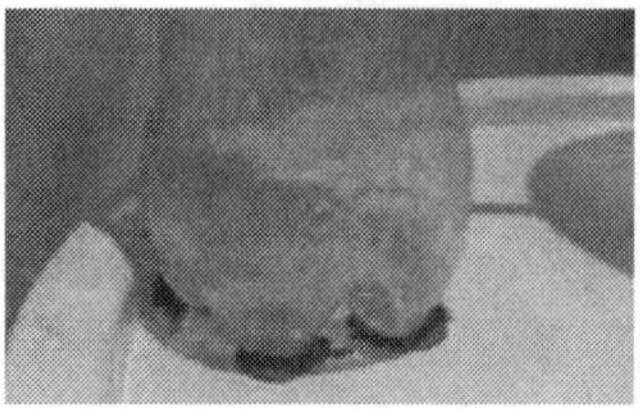 

Design Considerations:

- **Material Ratios:** Roughly 50% green plant material and 50% fresh manure yields best results.
- **System Type:** Static barrel for simple setups, digester with gas capture for advanced builds.

Building Process for Basic Bio-Fuel Production:

1. **Prepare the Materials:**
    - Chop or shred leaves into smaller pieces to speed decomposition.
    - Mix leaves and manure in a 1:1 ratio.
2. **Build the Digester:**
    - Clean and prepare a large, airtight drum or barrel.
    - Drill a hole in the lid for a hose fitting (if capturing gas).

    - Seal all joints with rubber gaskets and clamps.
3. **Load the Digester:**
    - Fill the container 2/3 full with the leaf and manure mixture.
    - Add enough water to create a thick slurry (like thin mud).
    - Seal the lid tightly.
4. **Fermentation Phase:**
    - Place the barrel in a warm, sunny location (ideally 85°F–100°F).
    - Allow to ferment for 30–90 days, depending on temperature.
    - Agitate gently once a week to encourage even breakdown.
5. **Harvest Bio-fuel:**
    - For biogas systems: Capture gas output through attached hoses into a storage bladder.
    - For bio-slurry systems: Drain liquid digestate and dry it into compressed fuel cakes or use as a rich compost booster.

Optional Biogas Capture:

- Install a simple hose and bladder system.
- Use captured methane for cooking fires, heating greenhouses, or running small burners.

Safety Measures:

- Always vent barrels safely; gas buildup can cause dangerous pressure.
- Keep digesters away from living areas.
- Wear gloves when handling manure and fermented materials.

Testing and Adjustments:

- If the digester smells strongly of ammonia, add more leaves.
- If the mixture is too dry, add water.
- If the mixture is too wet, add more dry leaves or straw.

Integration Tip: Position digesters near animal pens, composting areas, or greenhouses for easy access to manure and heat recycling. Use finished bio-fuel to support cooking stations, greenhouse heating, or emergency backup energy systems.

Benefits:

- Converts organic waste into usable energy
- Reduces dependence on wood, propane, or gasoline
- Produces high-value compost as a secondary product
- Supports a regenerative, closed-loop homestead ecosystem

Sustainability Focus: Build digesters from reclaimed barrels, surplus hose, and salvaged plumbing parts. Turn unavoidable waste streams into valuable inputs, reinforcing your property's self-sufficiency and reducing environmental impact.

Chapter 37:
DIY Toilet Paper Production

Maintaining personal hygiene is a fundamental aspect of healthy, off-grid living. Yet, few supplies disappear faster during crises than toilet paper. Learning to create your own sustainable alternatives ensures comfort and cleanliness without relying on fragile supply chains.

This project outlines simple, effective methods for producing reusable or biodegradable toilet paper substitutes using local, natural, and repurposed materials.

Materials Needed:

- Soft, absorbent natural fibers (cotton, bamboo fabric, old t-shirts, flannel)
- Scissors or rotary cutter
- Sewing machine or hand-sewing supplies (optional)
- Plant leaves (mullein, lamb's ear, large smooth leaves) for natural alternatives
- Basket or waterproof container (for storing clean cloths)

- Small bin or bucket with lid (for storing used cloths if using reusables)

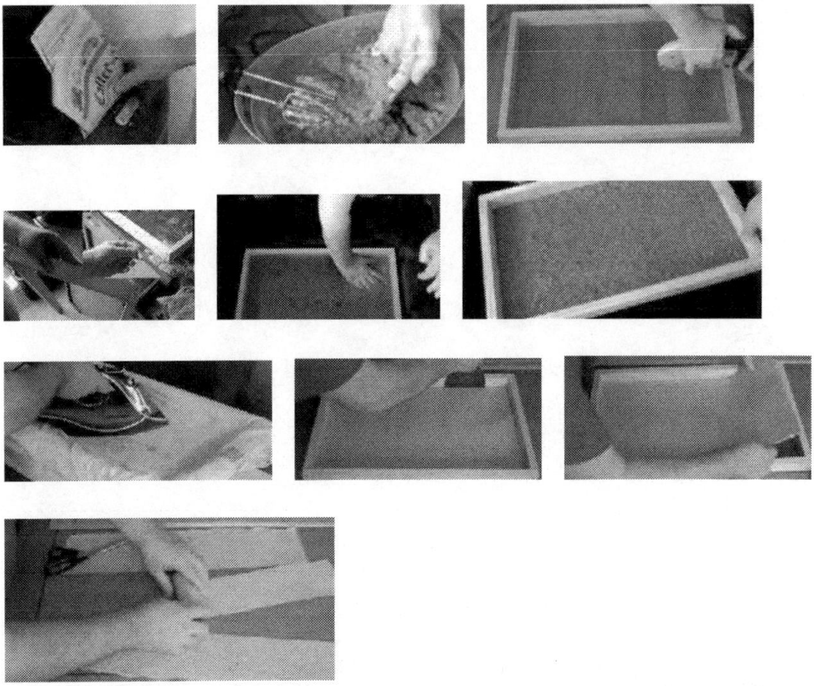

Design Considerations:

- **Reusable Cloths:** Washable and durable, ideal for long-term off-grid setups.
- **Biodegradable Plant Options:** Best for remote camping, outhouses, or emergency use.

Building Process for Reusable Toilet Paper:

1. **Select and Prepare Fabric:**
    - Choose soft, durable cloth.
    - Wash and dry fabric before cutting to remove residues and shrink it fully.
2. **Cut to Size:**
    - Cut fabric into manageable squares (6” x 8” or similar).
3. **Finish Edges (Optional):**
    - Sew around the edges with a zigzag stitch or serge to prevent fraying.
    - This step increases lifespan but is not strictly necessary.
4. **Organize and Store:**
    - Keep clean clothes in a waterproof basket or container.
    - Store used cloths in a sealed bin with mild soapy water until ready to wash.

Natural Plant-Based Alternatives:

1. **Identify Suitable Leaves:**
    - Mullein: Soft, large, and velvety
    - Lamb's ear: Thick and plush
    - Broadleaf plantain: Readily available and gentle

2. **Harvest Properly:**
   - Collect fresh, healthy leaves away from polluted areas.
   - Gently rinse and dry before use.
3. **Usage Tips:**
   - Use leaves fresh or store briefly in a cool, damp cloth.
   - Bury used leaves away from water sources to compost naturally.

Safety Measures:

- Avoid using unfamiliar plants; some can cause irritation.
- Wash reusable cloths thoroughly with hot water and natural disinfectants like vinegar.
- Sanitize storage bins regularly.

Integration Tip: Grow mullein, lamb's ear, and plantain around your home as part of your survival garden. This ensures a constant supply of biodegradable options, enhances pollinator habitats, and strengthens overall property resilience.

Benefits:

- Ensures long-term hygiene self-sufficiency

- Reduces dependency on external supply chains
- Supports sustainable waste management
- Enhances preparedness for emergencies and disruptions

Sustainability Focus: Use old clothing, linens, and towels that would otherwise go to waste. Compost plant-based

materials responsibly to close the loop, returning nutrients to the soil and minimizing environmental impact.

IV. Food Preparation, Preservation & Survival Cuisine

Off-grid living is not just about growing food — it's about preserving, preparing, and making it last through the seasons. Building a reliable food system strengthens your family's resilience, enhances nutrition, and creates real independence from vulnerable supply chains.

In this section, we'll cover critical techniques for:

- Preserving surplus harvests for the lean months ahead.
- Creating high-calorie, long-lasting survival rations.
- Cooking and preparing meals using low-tech, energy-efficient methods.

Each project emphasizes simplicity, sustainability, and integration with your broader off-grid systems — whether it's using solar ovens, homemade dehydrators, or naturally fermented preservation methods.

Mastering these skills transforms your property from a producer of food into a true fortress of food security, capable of withstanding droughts, disasters, and disruptions.

Let's dive into these essential off-grid kitchen and pantry practices, starting with an ancient technique for long-term egg preservation: water-glassing.

A. Food Preservation Techniques

Preserving your harvest is one of the most important skills in off-grid living. Without refrigeration, you must rely on proven methods to extend the life of your food — preventing waste and ensuring year-round nourishment.

This section introduces simple, effective techniques for keeping eggs fresh for months, preparing shelf-stable meals, and using traditional methods that require no electricity. These foundational skills support every other aspect of your survival plan, from garden to table.

By learning to preserve food naturally, you gain control over your pantry, reduce dependency on store-bought goods, and ensure your household stays fed no matter the season.

Chapter 38: Water-glassing Eggs with Pickling Lime

Fresh eggs are a nutritional powerhouse, but without refrigeration, their shelf life can be short. Thankfully, early homesteaders developed simple, chemical-free preservation methods to store eggs for months without spoilage.

Water-glassing is one of the most effective traditional techniques, using pickling lime to create an airtight seal around the eggs. Done properly, it can keep eggs fresh and ready to use for 8–12 months — a vital skill for true food security.

Materials Needed:

- Fresh, clean, unwashed eggs (ideally less than 24 hours old)
- Food-grade hydrated lime (pickling lime)
- Clean, filtered water (non-chlorinated)
- Food-grade storage container (ceramic crock, glass jar, or food-safe plastic bucket)
- Measuring cups

- Wooden spoon or non-metallic stirring tool

Building Process:

1. **Prepare the Lime Solution:**
    - Mix 1 ounce (about 2 tablespoons) of pickling lime into 1 quart of clean, non- chlorinated water.
    - Stir thoroughly until fully dissolved.
2. **Inspect the Eggs:**
    - Use only clean, unwashed, uncracked eggs.
    - Do not use eggs with hairline cracks or damaged shells, as they will spoil.
3. **Load the Container:**
    - Place eggs gently into the storage container, pointy side down.

    - Pour the lime solution over the eggs until they are completely submerged.
    - Add more solution as needed to maintain full coverage.
4. **Seal and Store:**
    - Cover the container loosely (do not make it airtight).
    - Store in a cool, dark place (50-70°F ideal).

Usage:

- Rinse eggs thoroughly before use.
- Eggs preserved with water-glassing are best suited for baking, scrambling, or hard boiling.

Safety Measures:

- Always use food-grade pickling lime.
- Do not use eggs that float before water-glassing; they are already aging.
- Check stored eggs monthly for cracks or discoloration.

Integration Tip: Combine water-glassing with your rainwater catchment and indoor greenhouse systems. Use fresh eggs daily when available and preserved eggs during lean periods, ensuring steady nutrition year-round without refrigeration.

Benefits:

- Preserves eggs for up to a year without refrigeration
- Easy, inexpensive, and scalable
- Supports year-round protein access in an off-grid system
- Enhances overall food security and sustainability

Sustainability Focus: Use rainwater or filtered well water to prepare the lime solution. Select local eggs from your own flock or nearby farmers to minimize transport and packaging waste, closing the sustainability loop.

Chapter 39:
Preserving Eggs with Isinglass and Mineral Oil

Before the advent of refrigeration, preserving eggs was a seasonal necessity. Another traditional method used by early homesteaders was coating eggs with isinglass — a natural, water-based preservative made from fish bladder gelatin — or applying a thin layer of mineral oil. Both techniques seal the eggshell pores, preventing air from entering and spoiling the contents.

When applied correctly, these methods can extend the shelf life of fresh eggs for 6 to 12 months, making them essential skills for off-grid living.

Materials Needed:

- Fresh, clean, unwashed eggs (no cracks)
- Food-grade mineral oil (odorless and tasteless)
- Isinglass powder (optional, traditional method)
- Warm water (for isinglass solution)
- Shallow bowl or small brush
- Clean, food-grade storage container

(cardboard egg cartons, baskets, or shallow trays)

Preserving with Mineral Oil:

1. **Warm the Mineral Oil:**
    - Slightly warm the mineral oil until it becomes easy to spread (do not overheat).
2. **Prepare the Eggs:**
    - Ensure eggs are fresh and dry.
    - Do not wash or scrub eggs, as this removes the protective bloom.
3. **Apply the Oil Coating:**
    - Dip eggs individually into the warm oil or use a small brush to apply a thin, even layer over the entire shell.
4. **Storage:**
    - Place coated eggs into cartons or trays, pointy end down.
    - Store in a cool, dark, consistent-temperature environment.

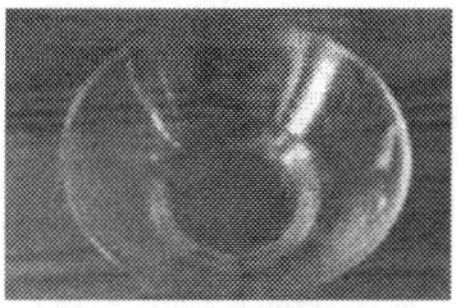

Preserving with Isinglass:

1. **Prepare the Isinglass Solution:**
   - Dissolve 1 ounce of isinglass powder in 1 quart of warm water.
   - Stir until fully dissolved.
2. **Load the Container:**
   - Place eggs gently into a storage container.
   - Pour the isinglass solution over the eggs, ensuring complete submersion.
3. **Seal and Store:**
   - Cover loosely and store in a cool, dark place.

Usage:

- Rinse eggs thoroughly before cooking.
- Suitable for all cooking applications once cleaned.

Safety Measures:

- Inspect eggs monthly; discard any with cracks, leaks, or unpleasant odors.
- Always handle eggs with clean, dry hands to prevent contamination.

Integration Tip: Combine mineral oil or isinglass preservation with a staggered egg collection schedule. This allows you to rotate stock efficiently and ensures a steady supply of preserved protein throughout winter months when hens lay less frequently.

Benefits:

- Extends egg shelf life for up to a year
- Requires minimal materials and equipment
- Simple, scalable, and historically proven
- Increases food security and off-grid sustainability

Sustainability Focus: Use minimal, natural materials — food-grade mineral oil or isinglass — to preserve eggs without modern refrigeration. Source eggs from your own flocks or barter networks, promoting local resilience and reducing reliance on commercial supply chains.

Chapter 40:

Meal-in-a-Bag: Hamburger Gravy and Mashed Potatoes Recipe

Preparing hearty, shelf-stable meals in advance is a crucial part of off-grid living and emergency readiness. "Meal-in- a-bag" recipes are compact, lightweight, and can be cooked with minimal resources, making them ideal for homesteaders, preppers, and survivalists.

This project details how to create a comforting, high- calorie meal of hamburger gravy and mashed potatoes — packaged for long-term storage and quick, easy preparation.

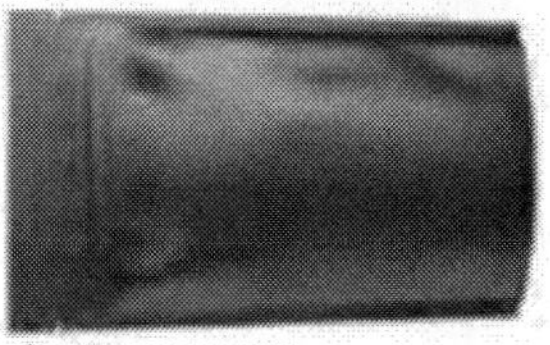

Materials Needed:

- Dehydrated ground beef (or freeze-dried beef crumbles)

- Dehydrated mashed potatoes (instant potato flakes)
- Dehydrated onions
- Dehydrated mushrooms (optional)
- Beef bouillon powder
- Cornstarch (for thickening)
- Salt, pepper, and desired spices
- Mylar bags (quart-size)
- Oxygen absorbers
- Vacuum sealer or impulse sealer
- Labels and permanent marker

Meal Assembly Process:

1. **Prepare Ingredients:**
   - Ensure all ingredients are fully dehydrated and free from moisture.
   - Measure single-serving or family-sized portions for each bag.
2. **Mix the Gravy Base:**
   - Combine dehydrated ground beef, onions, mushrooms (optional), beef bouillon powder, cornstarch, and seasonings in a bowl.
3. **Portion into Bags:**
   - Add one portion of the gravy base and one portion of dehydrated mashed

potatoes into each Mylar bag.
  - Keep mashed potatoes and gravy mix separated within the bag if possible (use small inner bags or divider pouches).

4. **Seal and Store:**
   - Insert an oxygen absorber.
   - Vacuum seal or heat seal the bag securely.
   - Label with meal name, date, and simple cooking instructions.

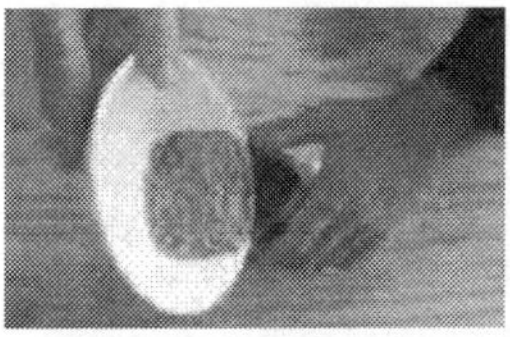
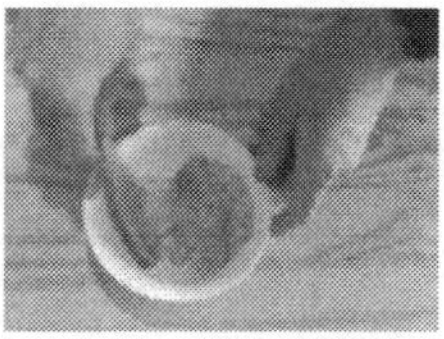
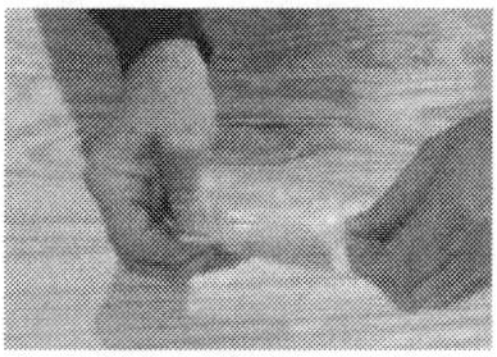
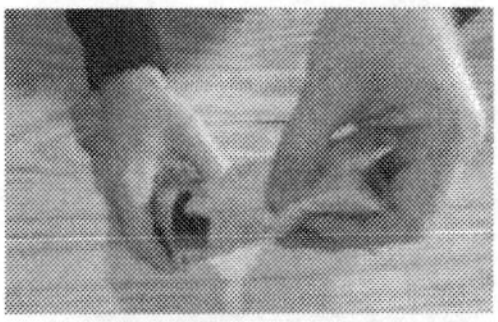

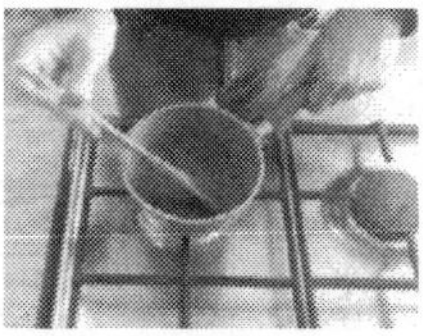

Cooking Instructions:

- Boil 2–3 cups of water.
- Add gravy mix to water, simmer for 5–8

minutes until thickened.

- Prepare mashed potatoes separately according to package instructions.
- Serve hot, layering mashed potatoes with rich hamburger gravy.

Safety Measures:

- Use only fully dehydrated, low-fat beef to prevent rancidity.
- Store meals in a cool, dark, dry place to maximize shelf life (up to 10 years if properly sealed).

Integration Tip: Use dehydrated ingredients from your home garden, greenhouse, or solar dehydrator projects to build your meal-in-a-bag pantry. Pair with rainwater catchment or off-grid heating systems for complete meal preparation independence.

Benefits:

- Quick, comforting meal requiring minimal resources
- Long-term storage capability
- High caloric density for physical labor and survival scenarios

- Modular: customize recipes based on available ingredients

Sustainability Focus: Dehydrate your own garden produce and grass-fed beef from local sources when possible.

Package meals in reusable or recyclable materials where practical to minimize environmental impact.

B. Long-Term Food Storage & Emergency Rations

In any self-reliant lifestyle, having reliable long-term food reserves is essential. Whether preparing for winter, power outages, or extended emergencies, preserved rations can mean the difference between comfort and crisis.

This section introduces practical, nutrient-dense food storage strategies designed for shelf life, portability, and ease of preparation. You'll learn how to dehydrate meat, make high-calorie ration bars, bake hard survival bread, and preserve essential fats and proteins — all without relying on refrigeration or modern supply chains.

Mastering these methods ensures you always have access to real, nourishing food when it matters most.

Chapter 41: How to Dry Meat and Turn It into Powder

Protein preservation is vital for sustaining strength and health during long-term off-grid living. Drying meat and grinding it into powder provides a lightweight, high-protein food that stores compactly and lasts for months or even years when done correctly.

This project walks you through drying and powdering meat using simple, sustainable methods that integrate beautifully with your solar dehydrator, food storage, and emergency ration systems.

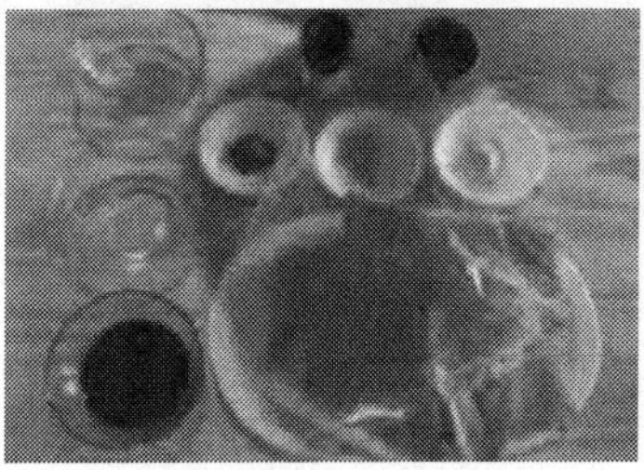

Materials Needed:

- Lean cuts of meat (beef, venison, rabbit, or poultry)
- Sharp knife and cutting board

- Salt or curing mix (optional but extends shelf life)
- Solar dehydrator, DIY drying racks, or a low- temperature oven
- Food processor, blender, or mortar and pestle (for powdering)
- Airtight containers, Mylar bags, or vacuum-sealed jars
- Oxygen absorbers (for long-term storage)

Preparation and Drying Process:

1. **Prepare the Meat:**
   - Trim all fat, sinew, and connective tissue (fat spoils quickly).
   - Slice meat into very thin strips, no more than 1/8" thick.
   - Optional: lightly salt or cure meat overnight for extra preservation.
2. **Dry the Meat:**
   - Arrange strips evenly on dehydrator trays or DIY racks.
   - Dry at 120°F–140°F until meat is hard, brittle, and snaps when bent (can take 12–48 hours depending on method and conditions).
   - Ensure complete dryness to prevent spoilage.
3. **Cool and Rest:**
   - Allow dried meat to cool completely before processing.

Powdering the Dried Meat:

1. **Break into Pieces:**

    - Snap dried strips into smaller chunks.
2. **Grind into Powder:**
    - Use a strong blender, food processor, or mortar and pestle to grind meat into a fine powder.
3. **Sift and Regrind (Optional):**
    - Sift powder through a fine mesh strainer.
    - Regrind any coarse bits for consistency.

Storage:

- Package meat powder into airtight containers with oxygen absorbers.
- Store in a cool, dark place.
- Shelf life: 1–2 years if properly dehydrated and sealed.

Usage:

- Add to soups, stews, gravies, or emergency ration bars for protein boost.
- Rehydrate with water and seasonings to create quick meat sauces.

- Mix with dried vegetables and grains for complete survival meals.

Safety Measures:

- Always use only lean, fresh meat.
- Ensure total dryness to prevent bacterial growth.
- Monitor storage for moisture or off smells.

Integration Tip: Combine powdered meat with other dehydrated ingredients from your pantry (like instant rice, dried beans, and vegetable flakes) to create "instant meal kits" for rapid, nutritious off-grid cooking.

Benefits:

- Extremely lightweight, portable protein source
- Highly compact for space-saving storage
- Versatile ingredient for multiple meals
- Enhances long-term food security and nutritional resilience

Sustainability Focus: Use wild game, responsibly raised livestock, or bartered meats to minimize reliance on commercial food systems. Dehydrate using solar energy

whenever possible, and package using reusable or recyclable materials.

Chapter 42:
2100 Calorie Emergency Ration Bars

In a survival situation, having compact, high-calorie foods that provide energy, nutrition, and shelf stability is critical. 2100 calorie emergency ration bars are dense, durable, and easy to prepare ahead of time, giving you a reliable source of sustenance in emergencies, bug-out scenarios, or during long periods off-grid.

This project teaches you how to make your own energy- rich survival bars with simple ingredients and minimal equipment, ensuring food security when it matters most.

Materials Needed:

- Rolled oats (2 cups)
- Powdered milk (1 cup)
- Sugar or honey (1/2 cup)
- Peanut butter or nut butter (1/2 cup)
- Water (1/2 cup)
- Optional: dried fruits, seeds, nuts, protein

powder, powdered vitamins

- Large mixing bowl
- Wooden spoon or sturdy spatula

- Baking sheet or mold
- Parchment paper
- Airtight bags or a vacuum sealer for storage

Preparation Process:

1. **Mix Dry Ingredients:**
    - In a large bowl, combine oats, powdered milk, sugar, and any optional dry ingredients (like seeds or protein powder).
2. **Prepare Wet Ingredients:**
    - Warm the peanut butter and water slightly to soften.
    - Mix thoroughly until smooth.
3. **Combine and Form Dough:**
    - Pour the wet mixture into the dry ingredients.
    - Stir until a thick, sticky dough forms.
4. **Shape the Bars:**
    - Press dough firmly into a lined baking sheet or mold.
    - Compact thoroughly to remove air

pockets (use a rolling pin or heavy object).

5. **Dry or Bake:**
    - Air-dry in a low-humidity environment for 1-2 days.
    - Alternatively, bake at 200°F for 1-2 hours until firm and dry.
6. **Cut and Package:**
    - Slice into individual bars (~400-600 calories each).
    - Vacuum seal or store in airtight bags with oxygen absorbers.

Nutritional Customization:

- Add powdered greens for vitamins.
- Mix in protein powder to boost protein content.
- Include dried fruits or nuts for variety.

Usage:

- Eat one bar per meal to sustain basic caloric needs in an emergency.
- Consume with water to aid digestion.

Safety Measures:

- Ensure complete dryness to prevent mold.
- Store in a cool, dry, dark place.
- Rotate bars every 6–12 months, depending on storage conditions.

Integration Tip: Stockpile emergency bars alongside your dehydrated meat powder, dried fruits, and survival bread. They form a core "grab-and-go" food kit that can be used for bug-out bags, vehicle emergency kits, and property emergency caches.

Benefits:

- High-calorie, nutrient-dense emergency food
- Easy to transport and ration
- Requires no refrigeration
- Extends survival capabilities during food shortages

Sustainability Focus: Use ingredients sourced from your own gardens, homestead dairy (for powdered milk), or local markets. Package bars in reusable vacuum bags or biodegradable wrappers whenever possible to minimize waste.

Chapter 43: How to Make Dandelion Bread & Hardtack (Emergency Survival Bread)

Bread has been a cornerstone of survival diets for centuries. When you need a nutrient-rich, shelf-stable food made from wild or readily available ingredients, dandelion bread and hardtack stand out as reliable options.

Dandelion bread taps into the abundant, nutritious greens often dismissed as weeds, while hardtack provides a virtually indestructible food source ideal for long-term storage and emergency reserves.

This project teaches you how to make both, ensuring your off-grid pantry includes durable, dependable survival staples.

Materials Needed:

- Fresh dandelion greens and blossoms (washed and chopped)

- All-purpose flour or whole wheat flour
- Baking powder or yeast (optional, for dandelion bread)
- Salt

- Water
- Mixing bowl
- Baking sheet
- Parchment paper
- Knife and measuring tools

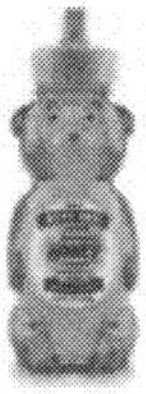

How to Make Dandelion Bread:

1. **Prepare the Ingredients:**
   - 2 cups flour
   - 1 cup chopped dandelion greens and petals
   - 1 tablespoon baking powder (or 1 teaspoon yeast for a slower rise)
   - 1/2 teaspoon salt
   - 1 cup water (adjust as needed for dough consistency)
2. **Mix the Dough:**
   - Combine dry ingredients first.
   - Stir in water and mix until a sticky dough forms.
3. **Shape and Bake:**
   - Form into a loaf or flatten into rounds.
   - Bake at 350°F for 25–35 minutes, until golden and firm.
4. **Cool and Store:**
   - Allow bread to cool fully.
   - Store wrapped in cloth or paper in a cool, dry area for short-term use.

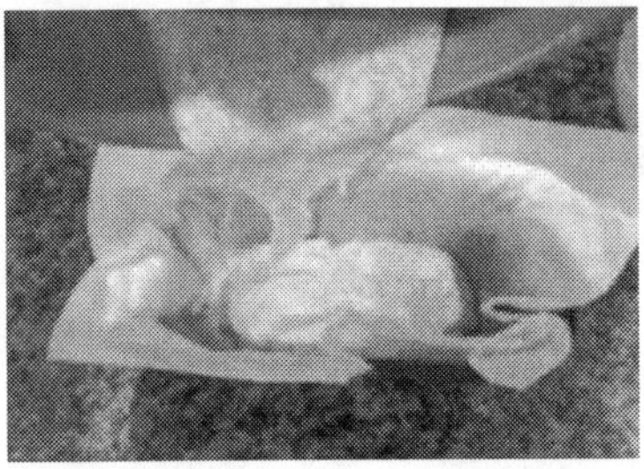

How to Make Hardtack (Emergency Survival Bread):

1. **Prepare the Ingredients:**
    - 3 cups flour
    - 1 cup water
    - 2 teaspoons salt
2. **Mix the Dough:**
    - Combine flour and salt.
    - Gradually add water, forming a stiff,

dry dough.

3. **Roll and Cut:**
    - Roll dough out to about 1/2-inch thickness.
    - Cut into squares (2" x 2") and pierce each square several times with a fork.
4. **Bake:**
    - Bake at 250°F for 2 hours, flipping halfway.
    - Allow to cool completely in the oven (this ensures maximum dryness).
5. **Storage:**
    - Store hardtack in airtight containers or vacuum-sealed bags.
    - Properly dried hardtack can last for years.

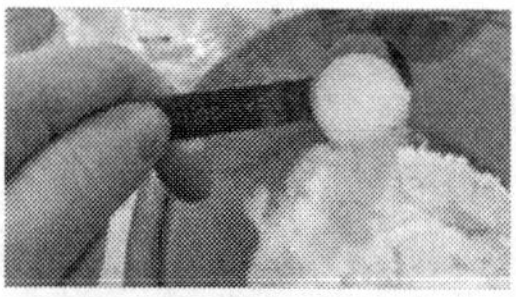

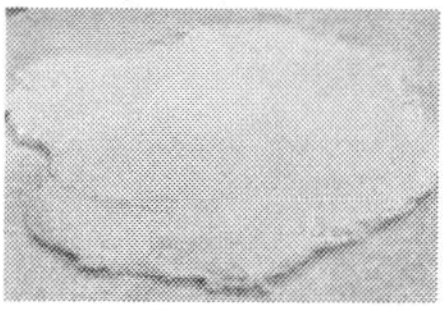

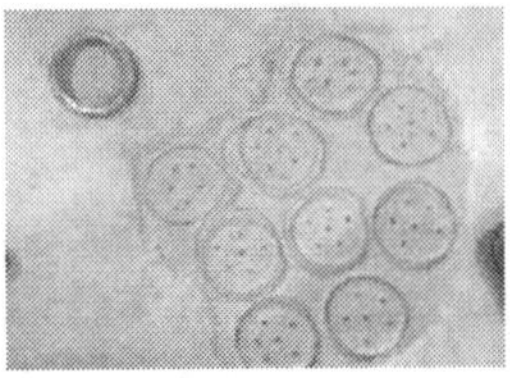
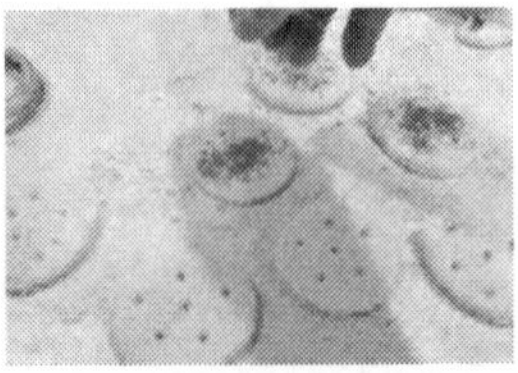
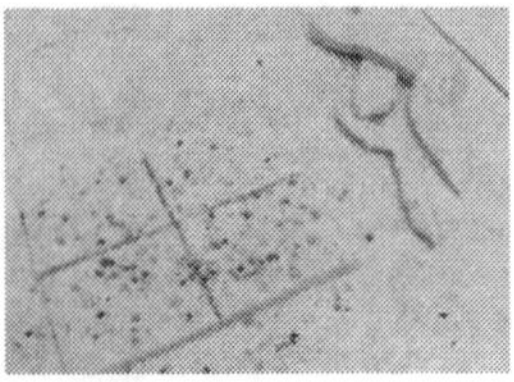
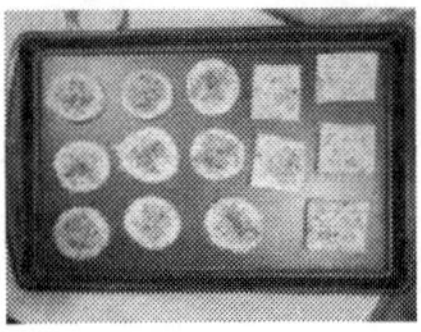

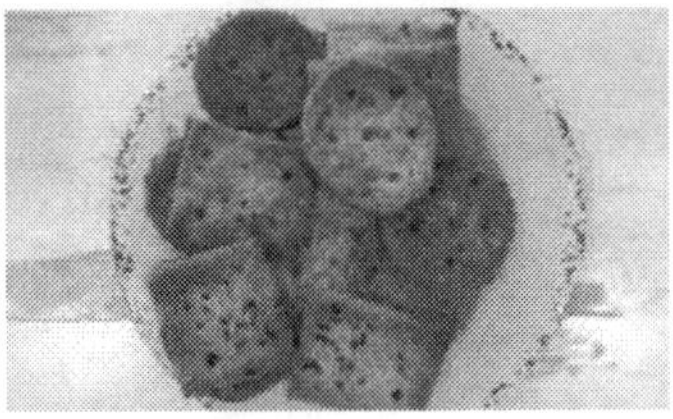

Usage:

- Eat dandelion bread fresh within a few days or toast slices to extend shelf life.
- Hardtack can be soaked in soups, stews, or water to soften before eating.

Safety Measures:

- Always identify dandelions properly to avoid toxic look-alikes.
- Ensure thorough drying and baking to prevent mold.

Integration Tip: Grow dandelions intentionally in a controlled patch alongside your other survival crops. Store hardtack with your dehydrated meats and ration bars to create diverse, long-lasting meal options during emergencies.

Benefits:

- Provides nutrient-dense and calorie-rich survival food
- Utilizes abundant natural resources
- Extremely shelf-stable and easy to prepare

- Adds variety to emergency food supplies

Sustainability Focus: Harvest wild or homegrown dandelions without chemicals. Use whole grains where possible to boost nutrition, and bake using solar ovens or rocket stoves to conserve fuel and energy.

Chapter 44:
Making Lard and Pemmican

When it comes to long-term, high-energy survival foods, few traditional methods surpass the practicality of rendering lard and preparing pemmican. Both were vital to early pioneers, explorers, and indigenous peoples for preserving calories, nutrition, and strength during harsh seasons.

Lard provides a versatile, stable fat for cooking and baking, while pemmican combines fat, dried meat, and sometimes berries into an ultra-dense, portable food ideal for emergency survival and off-grid sustenance.

This project teaches you how to produce both using sustainable, traditional techniques.

Materials Needed for Lard Rendering:

- Pork fat (leaf fat preferred) or beef suet
- Sharp knife
- Large pot or slow cooker
- Fine mesh strainer or cheesecloth
- Canning jars or metal tins for storage

Materials Needed for Pemmican:

- Dried, powdered lean meat (beef, venison, or bison)
- Rendered lard or tallow
- Dried berries (optional: blueberries, cranberries)
- Honey (optional, for sweetened pemmican)
- Mixing bowl
- Spoon or spatula
- Wax paper and airtight containers

Meat Preparation:

- Use previously dehydrated, powdered meat (see Chapter 41).
- Ensure it is completely dry to prevent spoilage.

Rendering Lard:

1. **Prepare the Fat:**
   - Trim fat into small cubes.
   - Remove as much meat and connective tissue as possible.
2. **Slow Heat the Fat:**
   - Place fat in a pot over very low heat.
   - Allow fat to melt slowly, stirring occasionally.
3. **Strain and Store:**
   - When bubbling slows and solids sink, strain liquid fat through cheesecloth.
   - Pour hot lard into sterilized jars.
   - Seal tightly and cool.

Making Pemmican:

1. **Mix Ingredients:**
   - In a large bowl, combine 2 parts powdered meat with 1 part melted lard.
   - Add finely chopped dried berries if desired.
   - Mix thoroughly into a thick paste.
2. **Shape and Cool:**
   - Form into balls, bars, or press into pans.
   - Let cool completely until firm.
3. **Wrap and Store:**
   - Wrap portions in wax paper.
   - Store in airtight containers in a cool, dark place.

Meat Preparation, Nutritional Value, and Cost Considerations:

- Pemmican is extremely energy-dense (around 700- 800 calories per 100 grams).

- High protein and fat content make it ideal for hard physical labor and survival.
- Producing pemmican at home can drastically reduce costs compared to buying commercial survival foods.

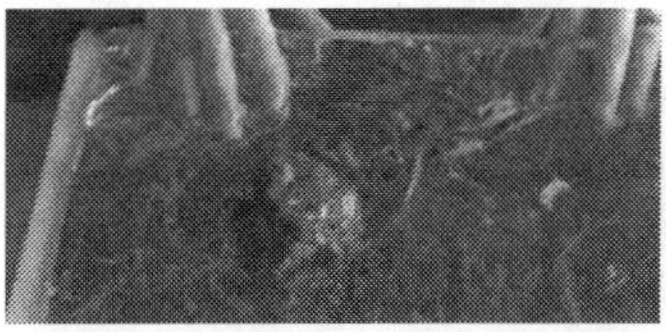
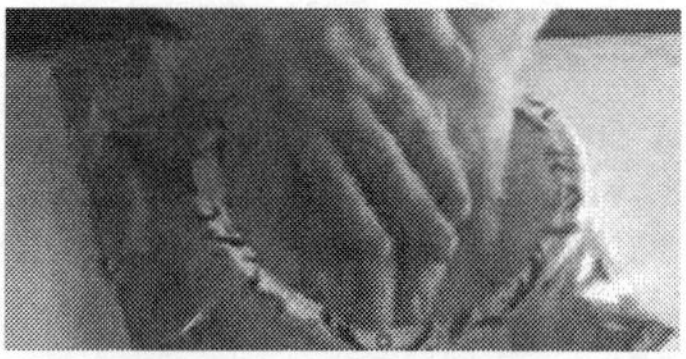

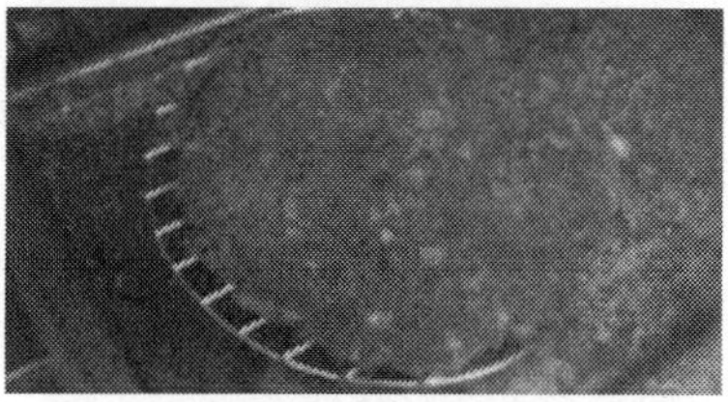

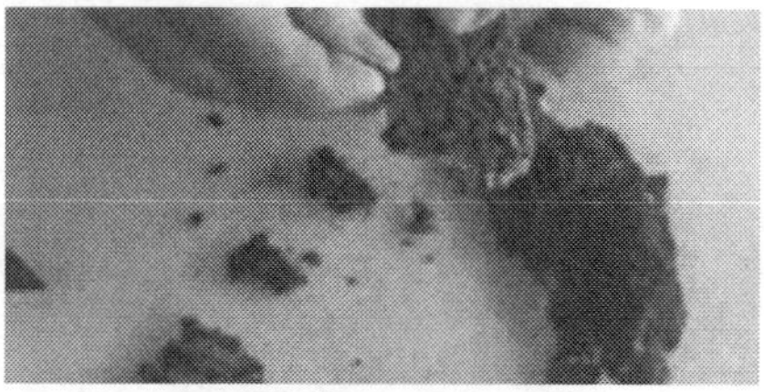

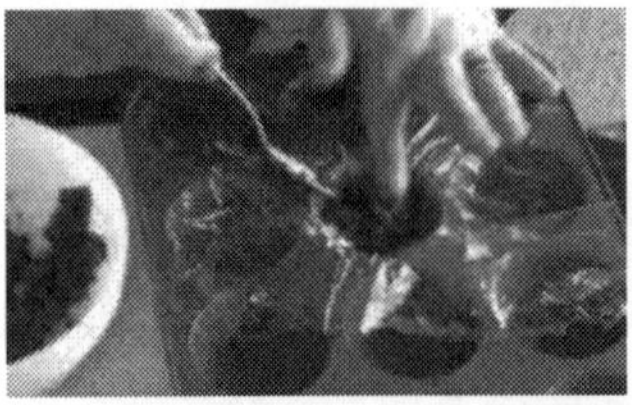
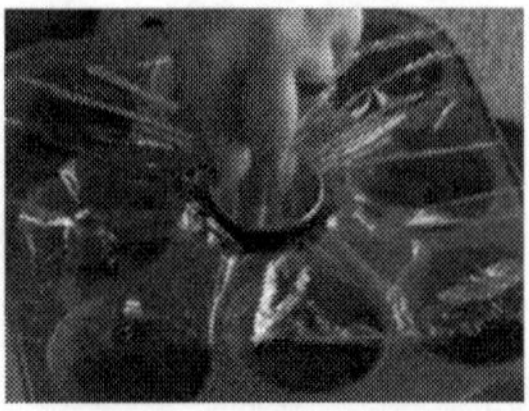
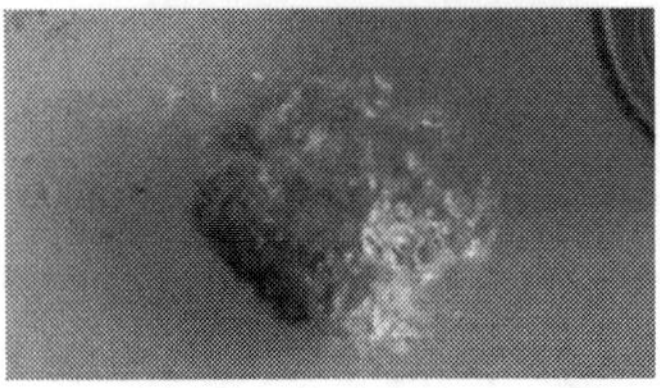

Safety Measures:

- Ensure all ingredients are completely dry to prevent mold.
- Render fat thoroughly to prevent rancidity.
- Rotate pemmican stock annually for maximum freshness.

Integration Tip: Pair pemmican with hardtack, dried fruits, and ration bars to create complete survival ration kits. Use homemade lard for everyday off-grid cooking, baking, and candle making, integrating fat preservation into multiple systems.

Benefits:

- Provides ultra-dense, long-lasting survival food
- Supplies essential fats and proteins

- Easily portable and durable
- Reduces waste by fully utilizing meat and fat resources

Sustainability Focus: Source animal fat and meat from your own livestock, local hunters, or ethical farms. Reuse glass jars and wax paper when possible to minimize packaging waste and environmental impact.

C. Hunting & Wild Game Acquisition

When living off-grid, securing a reliable source of protein becomes vital for survival and nutrition. Hunting and fishing are time-tested methods to supplement food stores, but they can be labor-intensive. Learning how to build and deploy automatic traps allows you to harvest wild game and fish passively, conserving valuable energy and resources.

This section teaches you simple, effective techniques for catching fish, small animals, and birds using sustainable, low-tech methods that can dramatically enhance your food resilience.

Chapter 45:
Automatic Traps for Fish (Gill Net and Trotline Assembly)

Fishing without constant attention or active labor dramatically increases your food security off-grid. Automatic traps like gill nets and trotlines allow you to passively harvest fish, conserving energy while providing a consistent protein source.

This project outlines the setup of basic, highly effective fish trapping systems using simple, sustainable materials.

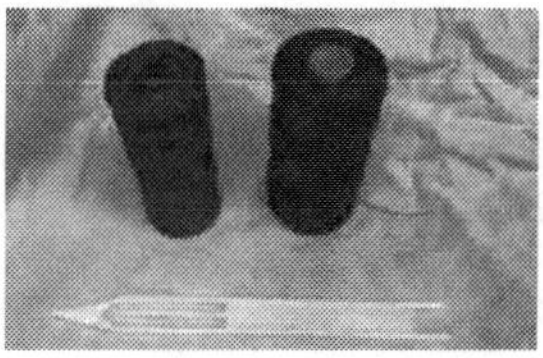  

Materials Needed:

- Gill net (pre-made or DIY from strong twine)
- Trotline rope (heavy-duty braided line)
- Fish hooks (barbed or barbless depending on regulations)
- Floats (plastic bottles, corks, or wood blocks)

- Weights (stones, scrap metal, sinkers)
- Knife and basic tools
- Gloves for handling rope and hooks

Building a Gill Net:

1. Prepare the Net
2. Attach Floats and Weights
3. Deploy the Net

Building a Trotline:

1. Prepare the Main Line
2. Attach Drop Lines
3. Anchor the Line
4. Bait and Deploy

Placement Strategies:

- Set gill nets in slow-moving water near fish pathways, like shallow banks or submerged logs.
- Deploy trotlines in deeper, flowing water where fish tend to feed or migrate.
- Use natural camouflage like reeds or debris to hide trap lines from sight.

Safety Measures:

- Clearly mark your trap locations with visible floats or tags to avoid accidental injury.
- Regularly inspect nets and lines to prevent entangling non-target species.
- Follow all local fishing regulations and ethical harvesting practices.

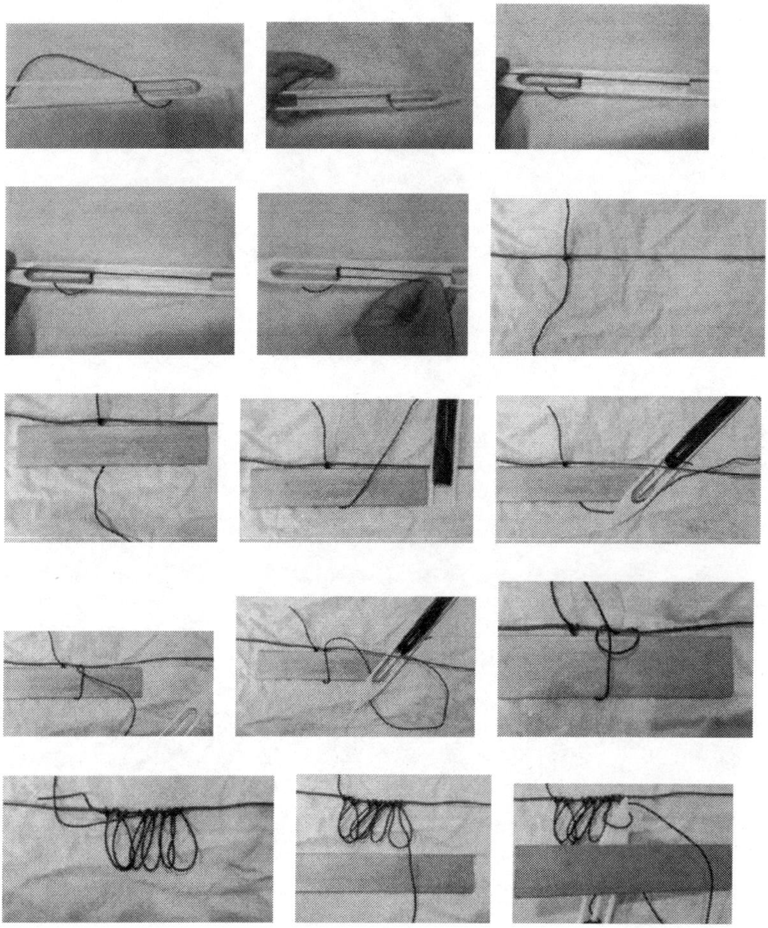

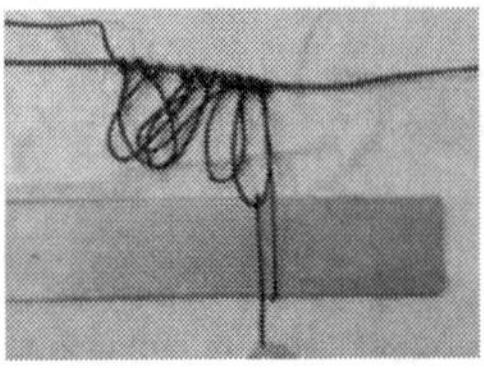
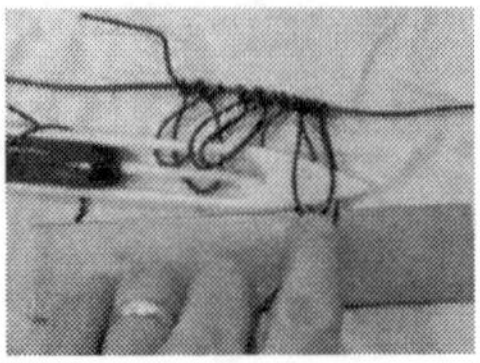
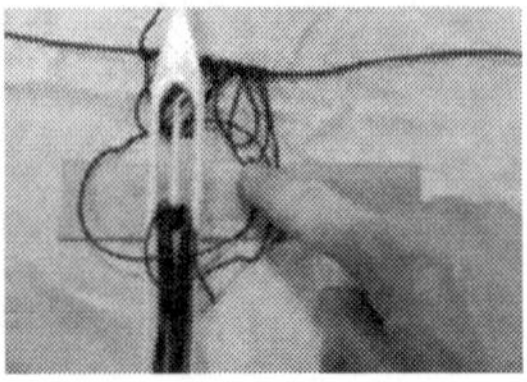
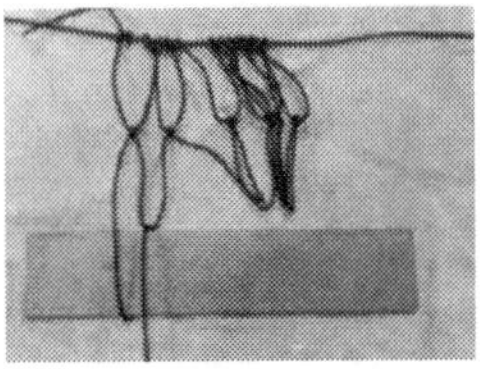
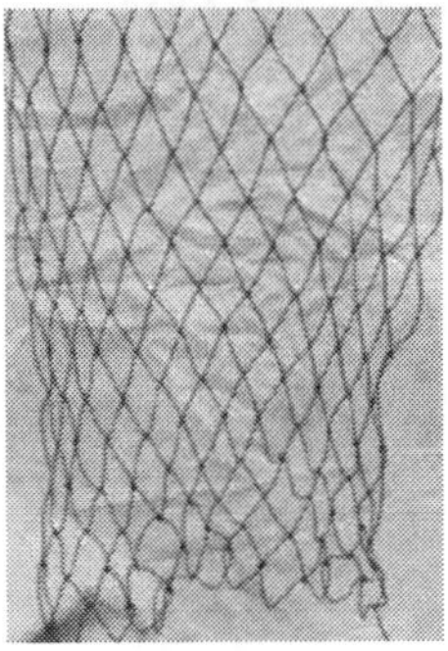

Trilene
XL
Trilene

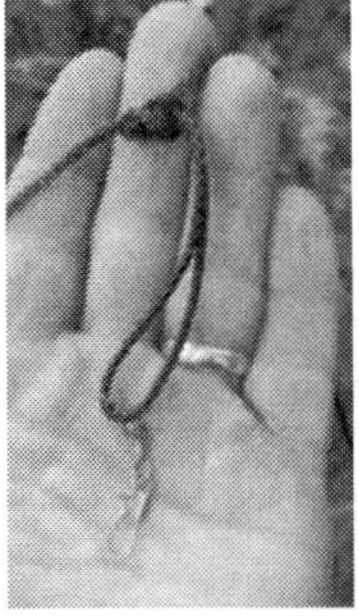
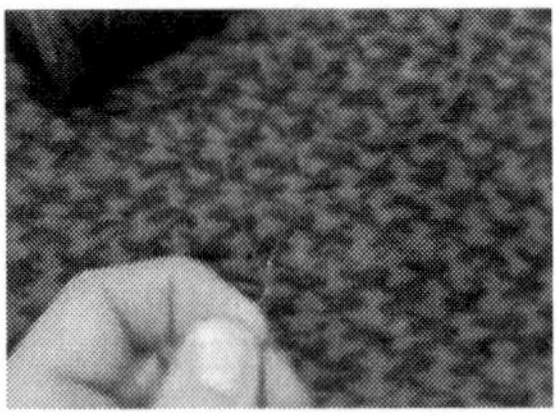

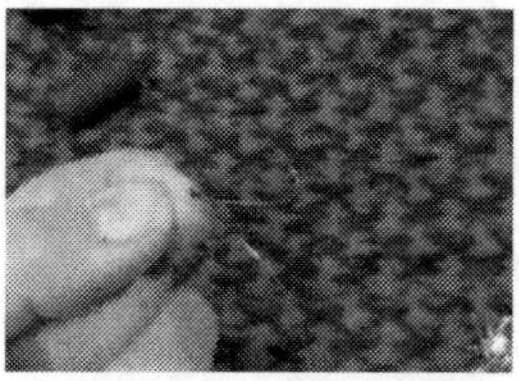
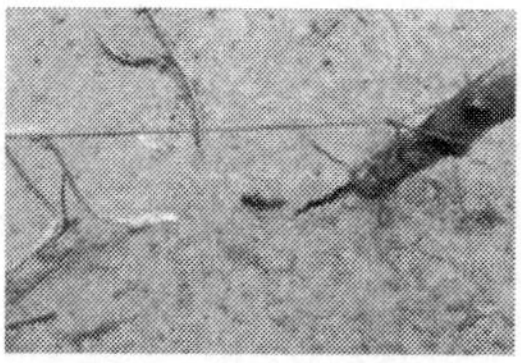
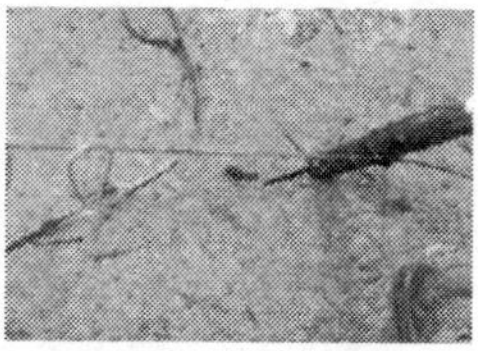

Integration Tip: Use fish caught through these traps to support your food preservation systems — drying, smoking, or fermenting — and feed nutrient-rich scraps to livestock, garden compost piles, or aquaponics setups.

Sustainability Focus: Craft nets and trotlines from natural or biodegradable materials where possible. Harvest only what you can consume to maintain local fish populations and ecosystem balance.

Chapter 46: Automatic Traps for Animals and Birds (DIY Wire Snare, Twitch-Up Snare, and Placement Strategies)

Introduction:

Automatic trapping for small animals and birds can dramatically supplement your food stores when hunting opportunities are limited. Wire snares and twitch-up traps allow you to harvest rabbits, squirrels, and ground-feeding birds using minimal equipment and energy.

Materials Needed:

- Strong, flexible wire
- Wooden stakes or branches
- Small sticks or saplings (for twitch-up triggers)
- Cutting pliers or multi-tool
- Gloves

DIY Wire Snare:

1. Prepare the Wire
2. Anchor the Snare
3. Position the Loop

Twitch-Up Snare Setup:

1. Prepare a Spring Pole
2. Set the Trigger
3. Connect the Snare

Placement Strategies:

- Place snares along established animal trails, near watering holes, or at tunnel entrances.
- Focus on natural funnels where terrain or brush directs animals into predictable paths.
- Use minimal disturbance to surroundings when setting traps to avoid alerting animals.

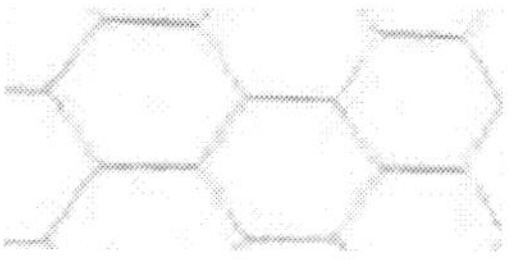

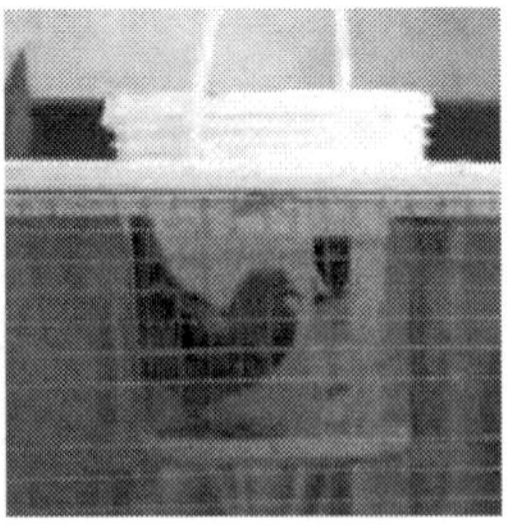

Safety Measures:

- Check traps at least twice a day to reduce animal suffering.
- Mark trap areas discreetly to protect yourself and others.
- Carry proper licenses or permits if required by local regulations.

Integration Tip: Combine trapping strategies with gardening systems: use captured small game as meat sources, and compost unused parts (bones, offal) to enrich garden soil.

Sustainability Focus: Harvest selectively and rotate trapping areas to prevent overhunting. Focus on abundant, fast-reproducing species and always respect the local ecological balance.

V. Gardening, Seeds, & Natural Remedies

In true off-grid living, food resilience begins with the soil beneath your feet. Beyond hunting and storage, building a thriving, self-renewing food production system through

gardens, seeds, and medicinal plants ensures independence for generations.

This section teaches you how to:

- Establish high-yield, sustainable gardens with minimal external inputs.
- Preserve and manage seed stock for long-term food security.
- Harness the healing power of natural remedies grown on your own land.

Each project not only sustains your pantry but strengthens your soil, your health, and your property's regenerative potential. By integrating gardens with water, compost, and energy systems, you create a closed-loop, living system that feeds, heals, and empowers.

Let's start by building a powerful food production engine: a DIY aquaponics system that raises both fish and vegetables together in a harmonious, water-efficient cycle.

Chapter 47: DIY Aquaponics System

Aquaponics is a revolutionary growing method that combines aquaculture (raising fish) with hydroponics (growing plants in water) into one efficient, self-sustaining system. Fish waste provides organic nutrients for plants, and the plants filter and clean the water for the fish, creating a perfect natural cycle.

Building your own DIY aquaponics setup means producing protein and vegetables year-round, using minimal land, water, and external fertilizers — an ideal solution for any off-grid homestead.

Materials Needed:

- Fish tank or large container (IBC tote, barrel, kiddie pool)
- Grow bed (plastic trays, troughs, or repurposed tubs)
- Substrate (clay pebbles, gravel, or coarse sand)
- Water pump (solar or DC-powered preferred)
- PVC pipes and fittings
- Aquarium air pump and air stones (optional but recommended)
- Net pots or planting cups
- Non-toxic water sealant (if using repurposed containers)
- Basic plumbing tools: drill, saw, pipe cutter, wrench
- Starter fish (tilapia, catfish, goldfish) and plant seeds or seedlings

Quick Setup Tips:

- **Fish:** Choose hardy, fast-growing species

suitable for your climate.

- **Plants:** Leafy greens, herbs, tomatoes, and peppers thrive in aquaponics systems.
- **System Type:** Start with a simple "flood and drain" (ebb and flow) system for reliability.

Building Process:

1. **Set Up the Fish Tank:**
    - Place tank in a shaded or semi-shaded area to prevent algae bloom.

- Fill with dechlorinated water and cycle for a few days.

2. **Assemble the Grow Bed:**
   - Position the grow bed slightly above the fish tank.
   - Fill with growing substrate.
3. **Install Plumbing:**
   - Connect a pump from the fish tank to the grow bed (inlet).
   - Install a bell siphon or simple drain system to return water to the fish tank.
4. **Cycle the System:**
   - Run the pump on a timer (15 minutes on, 45 minutes off).
   - Monitor water flow and drainage rates.
5. **Introduce Fish and Plants:**
   - Add fish once ammonia levels stabilize.
   - Plant seedlings directly into substrate.
6. **Monitor and Maintain:**
   - Check water pH, ammonia, and nitrate levels regularly.
   - Feed fish daily with appropriate feed.
   - Top off water as needed due to evaporation.

Materials, Costs, and PVC Assemblies:

- Use salvaged IBC totes, barrels, or bathtubs for tanks and beds.
- Basic setups can cost under $200 if repurposing materials.
- PVC pipes are inexpensive and easy to cut, shape, and seal.

Safety Measures:

- Ensure fish tanks are well-shaded to prevent overheating.
- Use only food-safe containers.
- Secure pumps and piping firmly to avoid leaks or failures.

Integration Tip: Connect aquaponics runoff into your outdoor gardens, using nutrient-rich water to fertilize soil beds. Integrate with rainwater catchment systems for full- cycle water reuse and greater efficiency.

Benefits:

- Produces fresh vegetables and fish together
- Uses 90% less water than traditional gardening
- Reduces dependence on chemical fertilizers
- Increases year-round food security in a

compact footprint

Sustainability Focus: Build your aquaponics system from reclaimed tanks, pipes, and containers. Source local fish breeds and heirloom seeds to reduce environmental footprint and enhance system resilience.

Chapter 48: Building a Mini Root Cellar

Long before refrigeration, root cellars provided reliable, low-energy storage for fruits, vegetables, and preserved foods year-round. Even a small, well-built root cellar can extend your harvests through winter, reduce waste, and strengthen your property's overall food resilience.

This project shows you how to build a compact, efficient root cellar using simple materials and techniques suited for off-grid living.

Materials Needed:

- Shovel or digging tools
- Insulated door or heavy wood panel
- Corrugated drainage pipe or PVC vent pipes (4-inch diameter)
- Gravel or crushed rock
- Cinder blocks, bricks, or salvaged stones
- Heavy tarp or waterproof membrane (optional for extra moisture control)
- Basic carpentry tools: saw, hammer, nails, screws
- Thermometer and humidity gauge

(optional but recommended)

Operational Mechanics:

- **Temperature:** A good root cellar stays between 32°F and 50°F.
- **Humidity:** Should be 85%–95% to prevent food dehydration.
- **Ventilation:** Essential to prevent mold buildup and control temperature.

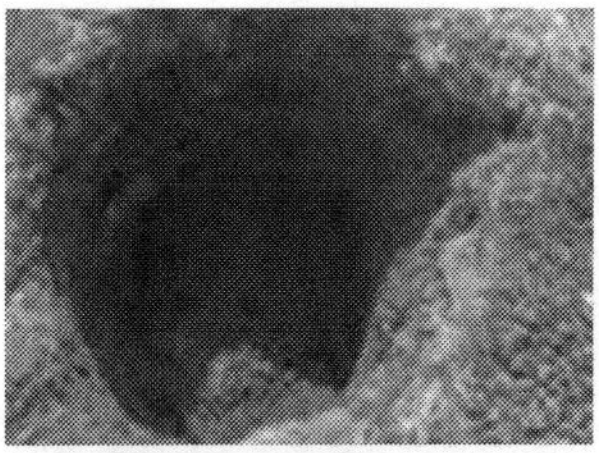

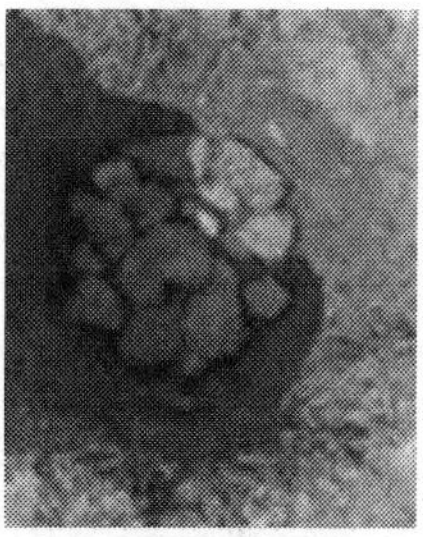

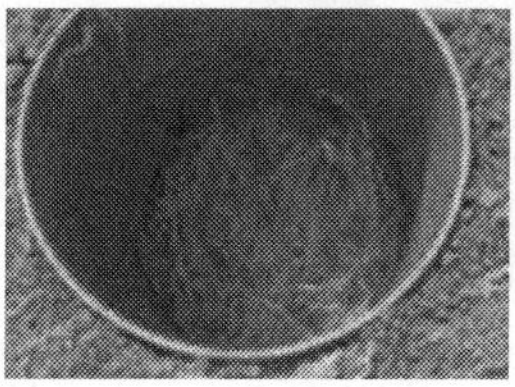

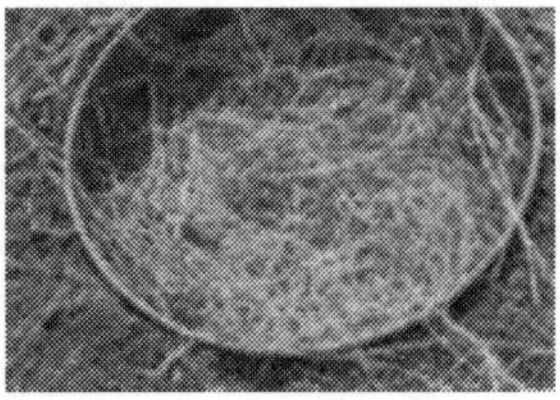

Building Process:

1. **Select the Site:**
    - Choose a shaded, north-facing slope if possible.
    - Well-drained soil is critical; avoid flood- prone areas.
2. **Excavate the Hole:**
    - Dig a hole about 4' to 6' deep and wide enough for your storage needs.
    - Remove large rocks and roots.
3. **Create the Floor:**
    - Spread 4–6 inches of gravel or crushed rock for drainage.

4. **Build the Walls:**
    - Line the sides with cinder blocks, stone, or brick for structural stability.
    - Leave small gaps or add simple weep holes at the bottom for water drainage if needed.
5. **Install Ventilation Pipes:**
    - Insert one intake pipe low near the floor (cold air entry).
    - Insert one exhaust pipe near the ceiling (warm air exit).
    - Cap pipes with mesh screens to keep out rodents and insects.
6. **Construct the Roof:**
    - Build a slanted wooden roof or use heavy salvaged materials.
    - Cover with waterproof membrane or heavy tarp for moisture protection.
    - Insulate well with soil, straw, or additional boards.
7. **Install the Door:**
    - Use a thick, insulated door to retain cool air.
    - Add weather stripping for a tight seal.

Food Storage Durations:

- Apples: 3–6 months
- Carrots: 4–6 months
- Potatoes: 5–8 months
- Cabbage: 3–5 months
- Onions: 5–7 months

Safety Measures:

- Ensure roof and walls are stable to prevent collapse.
- Monitor for mold, excess condensation, or rodents.
- Keep a thermometer inside to track conditions.

Integration Tip: Link your mini root cellar with your outdoor kitchen, greenhouse, and preservation zones. Use harvested rainwater for irrigation and cleaning, recycling runoff back into your gardens.

Benefits:

- Provides reliable, passive food storage
- Reduces dependency on refrigeration and external power
- Extends harvest life and minimizes waste

- Creates a vital backup food reserve for emergencies

Sustainability Focus: Use reclaimed materials (old bricks, scrap lumber, salvaged insulation) to build your root cellar.

Blend it naturally into your landscape to minimize environmental disturbance and visual impact.

Chapter 49: DIY Survival Garden & Wall- Hanging Herb Garden

A self-reliant homestead thrives on a steady, diverse food supply. Establishing a DIY survival garden and compact herb gardens ensures a constant flow of fresh vegetables, medicinal plants, and culinary herbs without relying on commercial seeds or grocery stores.

This project shows how to build a resilient, high-yield outdoor survival garden and a space-saving wall-hanging herb garden perfect for limited spaces, balconies, or indoor setups.

Materials Needed for Survival Garden:

- Quality heirloom seeds
- Hand tools: hoe, rake, spade, watering can
- Compost and natural fertilizers
- Mulch (straw, wood chips, leaves)

- Raised bed materials (optional: wood, stone, bricks)
- Fencing material (chicken wire, repurposed pallets) to protect from pests

Materials Needed for Wall-Hanging Herb Garden:

- Wooden pallet, shoe organizer, or vertical planter frames
- Landscape fabric or geotextile cloth
- Potting soil
- Small herb seedlings or seeds (basil, oregano, thyme, mint, cilantro, etc.)
- Staple gun and staples (if using pallets)
- Hooks or screws for mounting

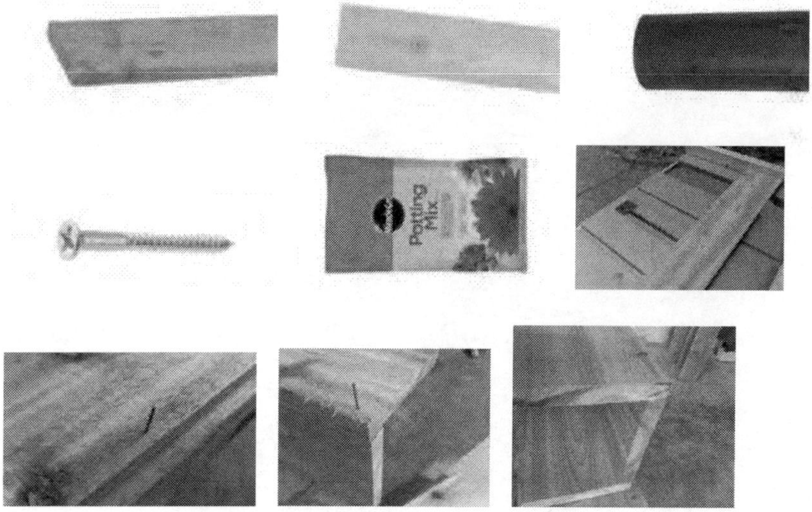

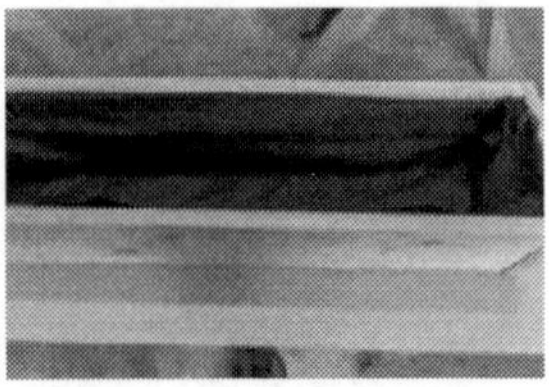

Design Considerations:

- **Survival Garden:** Prioritize calorie-dense, nutrient- rich, and fast-growing crops.
- **Wall-Hanging Garden:** Focus on herbs that can be continually harvested.

Building the Survival Garden:

1. **Choose a Location:**
    - Select a sunny spot with good drainage.
    - Clear rocks, weeds, and debris.
2. **Prepare the Soil:**
    - Loosen topsoil 8-12 inches deep.
    - Mix in compost or natural fertilizer.
3. **Plant Important Crops:**

- **Staples:** Potatoes, beans, corn, squash
- **Greens:** Kale, spinach, collards
- **Roots:** Carrots, beets, radishes
- **Medicinals:** Garlic, onions, calendula

4. **Apply Mulch:**
    - Mulch heavily to conserve moisture, suppress weeds, and enrich soil.
5. **Install Simple Protection:**
    - Build low fencing or netting to deter rabbits, deer, and birds.

Building the Wall-Hanging Herb Garden:

1. **Prepare the Structure:**
    - Clean and sand the pallet or frame.
    - Staple landscape fabric to create planting pockets.
2. **Fill and Plant:**
    - Fill pockets or planter slots with potting soil.
    - Plant herb seeds or seedlings.
3. **Mount the Garden:**
    - Hang securely on a wall that receives at least 4–6 hours of sunlight daily.
4. **Watering and Maintenance:**
    - Water regularly.

    - Trim herbs to encourage new growth.

Important Crops and Herb Usage:

- **Survival Crops:** Beans (protein), squash (storage), potatoes (calories)
- **Herbs:** Mint (digestion), basil (immune booster), thyme (respiratory health)

Safety Measures:

- Check pallet wood for chemical treatment stamps (use only "HT" heat-treated wood).
- Monitor gardens for pests and diseases naturally (use companion planting where possible).

Integration Tip: Link your survival garden to your rainwater catchment system for sustainable irrigation. Place your herb wall near your outdoor kitchen, solar dehydrator, or indoor greenhouse for easy culinary and medicinal access.

Benefits:

- Ensures constant access to fresh food and healing herbs
- Reduces reliance on external food supplies

- Supports year-round nutrition and health
- Optimizes small spaces with vertical growing techniques

Sustainability Focus: Save seeds each season to create a self-renewing food source. Reuse pallets, containers, and local soil amendments. Focus on perennial crops and regenerative practices to build soil health over time.

B. Seed Stockpiling & Soil Health

Healthy seeds and living soil form the heart of any successful off-grid homestead. Without strong seeds and fertile ground, even the most advanced systems will struggle to provide lasting food security.

Seed stockpiling ensures that you control your future harvests, saving varieties adapted to your land and climate. Maintaining soil health guarantees that those seeds grow into thriving, nutrient-dense crops for years to come.

In this section, you'll learn how to:

- Preserve a resilient seed bank tailored to your needs.
- Analyze and improve your soil using simple, low- cost methods.
- Build a regenerative system where soil fertility and food production strengthen each other year after year.

Mastering seeds and soil means mastering the foundation of your self-sufficiency journey.

Chapter 50:
How to Stockpile Seeds

Seeds are the ultimate form of food security. Having a diverse, healthy stockpile of seeds ensures that no matter the season or crisis, you can regrow your own supply of fruits, vegetables, herbs, and grains. Learning to properly select, store, and manage seeds transforms your garden from a one-time project into a permanent, renewable system.

This project guides you through building and maintaining a sustainable seed stockpile to secure your long-term food independence.

Materials Needed:

- Heirloom or open-pollinated seeds
- Airtight storage containers (glass jars, Mylar bags, vacuum-sealed bags)
- Silica gel packets or rice (for moisture control)
- Labels and permanent marker
- Seed envelopes or small paper bags
- Cool, dark, dry storage location (root cellar, basement, sealed container)

Best Seeds for Long-Term Survival and Shelf Life:

- Beans (green beans, lima, kidney)
- Corn (dent corn, sweet corn)
- Squash (winter and summer varieties)
- Tomatoes (paste and slicing varieties)
- Peppers (hot and sweet)
- Carrots, beets, radishes (for fast-growing root crops)
- Cabbage, kale, collards (cold-hardy greens)
- Onions and garlic (bulb-forming vegetables)
- Herbs: basil, oregano, cilantro, thyme

Building the Seed Stockpile:

1. **Choose High-Quality Seeds:**
   - Focus on heirloom and open-pollinated varieties to ensure you can save seeds year after year.
   - Avoid hybrid seeds (labeled "F1"), which do not produce true-to-type offspring.
2. **Package for Preservation:**
   - Place seeds in small paper envelopes first (allows seeds to breathe).
   - Label each packet with plant name, variety, and collection year.
3. **Seal and Store:**
   - Pack seed envelopes into airtight jars, Mylar bags, or vacuum-sealed pouches.
   - Add silica gel packets or dried rice to absorb moisture.

4. **Find the Right Storage Location:**
    - Ideal storage conditions are 35°F–50°F and low humidity.
    - Root cellars, cool basements, or solar-cooled sheds work well.
5. **Rotate and Test:**
    - Every 2–3 years, test a few seeds by germinating them to ensure viability.
    - Replace or replenish stock as needed.

Seed Shelf Life (General Guidelines):

- Beans: 3–5 years
- Corn: 1–3 years
- Squash: 4–6 years
- Tomatoes: 4–6 years
- Peppers: 2–5 years
- Carrots: 2–4 years

Safety Measures:

- Keep seed storage areas free from rodents and insects.
- Avoid exposing seeds to sunlight, moisture, or frequent temperature swings.

Integration Tip: Pair seed saving and stockpiling with your survival garden system. After each harvest, collect seeds from your healthiest plants to continuously strengthen your local seed genetics and enhance food resilience.

Benefits:

- Guarantees future planting cycles without external resupply
- Saves money year after year
- Preserves rare and valuable plant varieties
- Increases self-sufficiency and food system resilience

Sustainability Focus: Save seeds from your own crops and share with neighbors to create community seed banks.
Focus on varieties adapted to your specific growing conditions for maximum success and environmental harmony.

Chapter 51:
DIY Mason Jar Soil Test

Good soil is the foundation of every thriving garden and homestead. Before you plant your first seed or transplant your first seedling, it's essential to know what's happening beneath your feet. A DIY mason jar soil test is a simple, low-cost method to understand your soil's basic composition—whether it's heavy clay, loose sand, or rich loam.

Knowing your soil type empowers you to amend, enrich, and manage it wisely, increasing your yields and building long-term fertility without synthetic inputs.

Materials Needed:

- Mason jar with lid (quart-size preferred)
- Shovel or soil trowel
- Permanent marker
- Ruler or measuring tape
- Dish soap
- Water
- White vinegar and baking soda (optional for pH testing)

Understanding the Base Components of Soil:

- **Sand:** Large, gritty particles that drain water quickly.
- **Silt:** Smooth, flour-like particles that retain water moderately.
- **Clay:** Tiny, sticky particles that hold water tightly.
- **Organic matter:** Decayed plant and animal material that feeds your soil.

   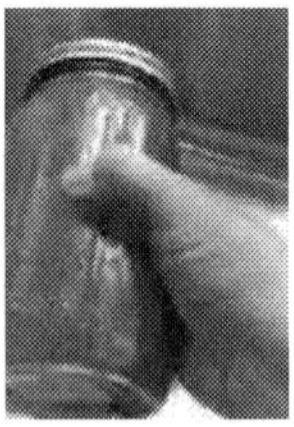

Testing Process:

1. **Collect Soil Sample:**
   - Dig 6–8 inches deep in your garden.
   - Remove any rocks, roots, or debris.
   - Fill the mason jar halfway with loose soil.
2. **Fill with Water:**
   - Add water to nearly the top of the jar.
   - Add 1 teaspoon of dish soap (helps separate soil particles).
3. **Shake Vigorously:**
   - Seal the lid tightly.
   - Shake for 3–5 minutes until soil and water are fully mixed.
4. **Let Settle:**
   - Place the jar on a flat surface.
   - Allow it to settle undisturbed for 24–48 hours.

5. **Analyze the Layers:**
    - Sand will settle at the bottom (first few hours).
    - Silt forms the next layer (after several more hours).
    - Clay sits at the top (after full settling).
6. **Measure and Record:**
    - Use a ruler to measure each layer's depth.
    - Calculate percentages to understand your soil composition.

Optional Quick pH Test:

- Mix a spoonful of moist soil with white vinegar (bubbles = alkaline soil).
- Mix a separate spoonful of moist soil with water and baking soda (bubbles = acidic soil).

Soil Improvement Tips Based on Results:

- **Sandy Soil:** Add compost and organic matter to improve water retention.
- **Clay Soil:** Add sand, compost, and mulch to enhance drainage.
- **Silty Soil:** Focus on organic matter and gentle tilling to prevent compaction.

Safety Measures:

- Avoid sampling near driveways, treated wood, or contaminated areas.
- Always wash hands after handling soil samples.

Integration Tip: Test your soil annually alongside composting and cover cropping routines. Monitor improvements over time to optimize planting strategies and ensure robust harvests.

Benefits:

- Provides critical insight into soil health with minimal tools
- Guides smart soil amendments and planting decisions
- Strengthens garden productivity naturally
- Supports long-term regenerative practices

Sustainability Focus: Use household materials for testing. Focus soil amendments on natural, local inputs such as compost, leaf mulch, and green manure crops to close fertility loops and avoid synthetic chemicals.

Chapter 52: DIY Projects on Seeds, Herbs, and Natural Remedies

A resilient off-grid lifestyle doesn't just depend on food production — it also hinges on maintaining health naturally. By learning how to grow, prepare, and apply herbal remedies, you create a living pharmacy on your property, capable of treating common ailments and boosting overall wellness without dependency on commercial drugs.

This project teaches you how to make simple, powerful natural remedies using homegrown or wildcrafted plants, helping you prepare for everyday needs and unexpected emergencies.

Materials Needed:

- Fresh cabbage leaves
- Moringa leaves (fresh or dried)
- Basic kitchen tools: knives, cutting boards, blender or mortar and pestle
- Food dehydrator or solar dehydrator

- Cheesecloth or muslin cloth
- Airtight jars or containers
- Labels and markers
- Basic first-aid supplies (for assembling kits)

Making Cabbage Bandages for Inflammation and Joint Pain:

1. **Select Fresh Leaves:**
    - Choose large, clean cabbage leaves.
2. **Prepare the Leaves:**
    - Remove the thickest veins.
    - Lightly bruise the leaves with a rolling pin to release healing juices.
3. **Apply as a Bandage:**
    - Wrap cabbage leaves around sore joints or inflamed areas.
    - Cover with cloth or gauze and leave on for 1–3 hours.
4. **Storage:**
    - Store fresh leaves in a cool place or refrigerate for short-term use.

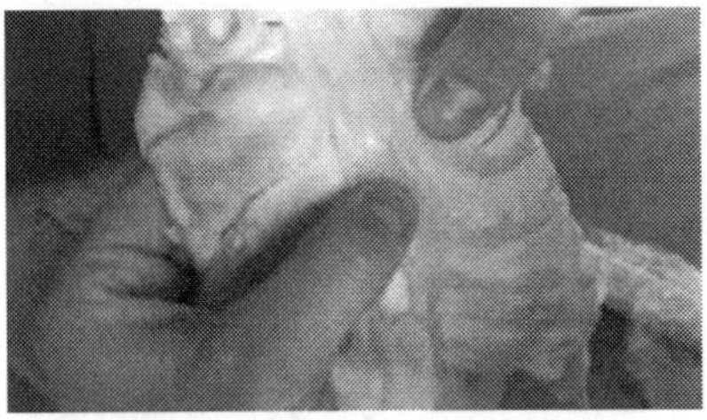
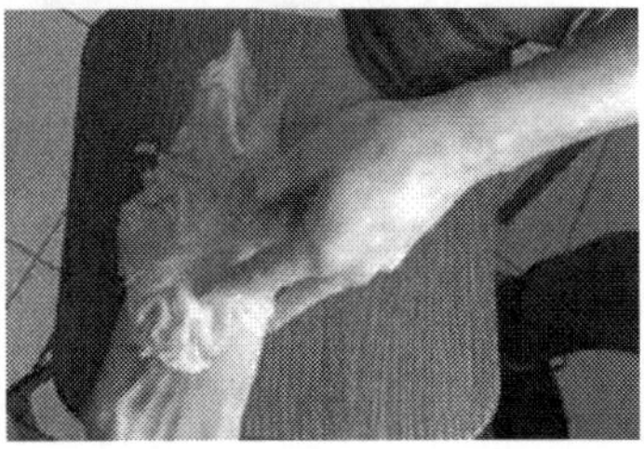
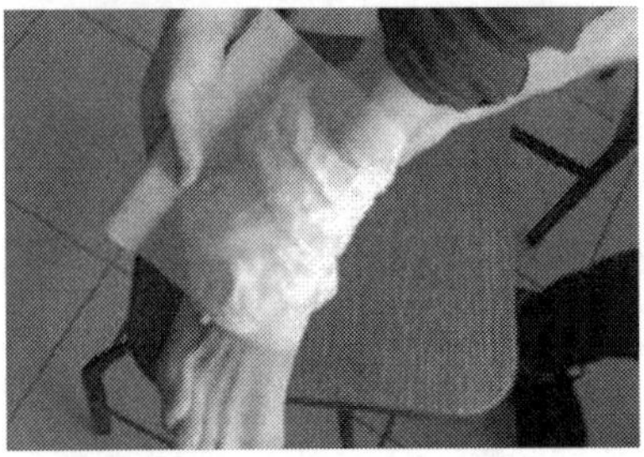
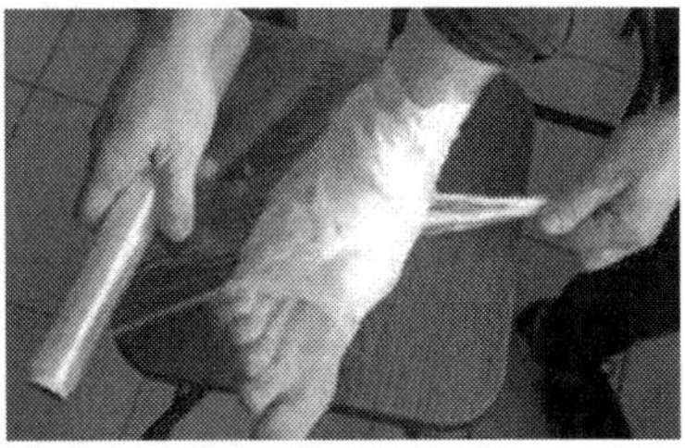
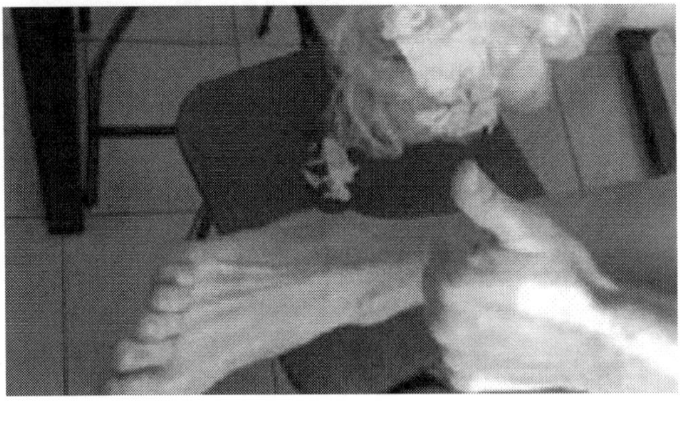

How to Make Moringa Powder:

1. **Harvest and Dry:**
   - Pick healthy moringa leaves.
   - Dehydrate leaves using a solar dehydrator or a well-ventilated, shaded area.
2. **Grind into Powder:**
   - Crush dried leaves into a fine powder using a blender or mortar and pestle.
3. **Store Properly:**
   - Place powder in airtight jars.
   - Store in a cool, dry, dark place.
4. **Usage:**
   - Add to smoothies, soups, or sprinkle onto meals for a major nutritional boost.

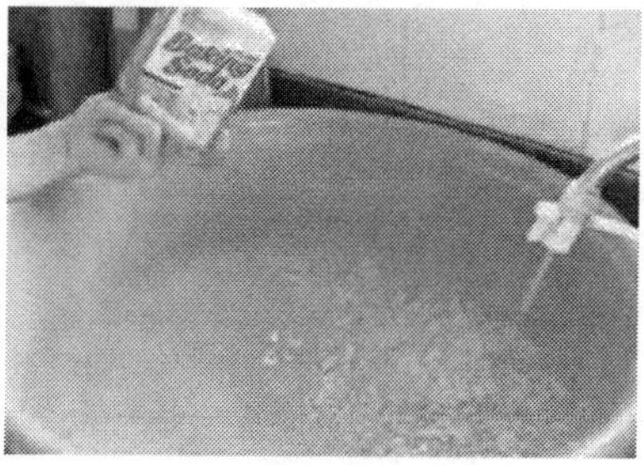

Crafting a SHTF Medicinal Kit:

Organize natural remedies by body system for faster use during emergencies:

- **Circulatory:** Hawthorn, garlic (support heart health)
- **Digestive:** Peppermint, ginger (ease upset stomach)
- **Endocrine:** Licorice root (supports adrenal health)
- **Integumentary (Skin):** Calendula, aloe vera (heal wounds, burns)
- **Immune:** Echinacea, elderberry (boost immune response)
- **Muscular:** Arnica (relieve muscle aches)
- **Nervous:** Valerian root, chamomile (calm anxiety, promote sleep)

- **Renal:** Dandelion root (support kidney health)
- **Reproductive:** Raspberry leaf (support women's health)
- **Respiratory:** Mullein, thyme (support lung function)

Safety Measures:

- Positively identify all herbs before harvesting or using.
- Consult reliable herbal resources and dosing guidelines.
- Monitor for allergic reactions with new remedies.

Integration Tip: Grow medicinal plants alongside vegetables in your survival garden. Designate specific raised beds or sections for herbs, rotating crops each year to enhance soil health and biodiversity.

Benefits:

- Reduces dependency on pharmaceutical supply chains
- Empowers proactive natural health care
- Enhances property self-sufficiency and resilience
- Creates multi-purpose gardens with food and medicine

Sustainability Focus: Use heirloom seeds and local plant varieties. Preserve herbs using natural dehydration methods and store remedies in reusable glass jars or compostable packaging to minimize environmental impact.

C.Natural Remedies & Medicinal Plants

Nature has always been humanity's first pharmacy. Long before modern medicine, people relied on the healing power of plants to treat wounds, cure illnesses, and strengthen the body against disease.

In an off-grid world where access to commercial healthcare may be limited, building your own natural remedy toolkit is not just wise — it's essential.

This section shows you how to:

- Grow powerful medicinal herbs right alongside your food crops.
- Prepare simple, effective remedies from common plants.
- Build a natural medicine cabinet that supports your health year-round.

By mastering these skills, you restore a vital form of independence: the ability to care for yourself and your family with the gifts already growing on your land.

Let's begin reclaiming the healing traditions that

have sustained human life for centuries.

Chapter 53:
The Only 7 Seeds You Need to Stockpile for a Crisis

When space, resources, or time are limited, focusing on a core set of seeds ensures that you can maintain a complete, resilient food supply through any crisis. These "critical seven" crops provide essential calories, nutrition, and seed- saving potential, allowing you to sustain yourself year after year.

This project outlines the most valuable survival seeds to prioritize and explains why they are essential for off-grid self-reliance.

Staple Crops: Nutrient-Dense Root Vegetables, Grains, and Greens

1. **Beans (Protein Powerhouse):**
    - High in protein, fiber, and calories.
    - Excellent nitrogen fixer to improve soil fertility.
    - Easy to dry, store, and replant.
2. **Corn (Calorie Base):**

- Dense in carbohydrates and versatile for food, animal feed, and even fuel.
- Dent corn varieties are best for grinding into meal.

3. **Squash (Storage Champion):**
   - Winter squash stores for months without refrigeration.
   - High in vitamins A and C.
4. **Potatoes (Reliable Starch Source):**
   - High yield per square foot.
   - Grows well even in marginal soils.
5. **Carrots (Nutrient-Dense Root Vegetable):**
   - Rich in beta-carotene and vital minerals.
   - Stores well when properly cellared.
6. **Kale (Hardy Leafy Green):**
   - Survives frost and provides vital vitamins throughout winter.
   - Continuous harvest when properly managed.
7. **Amaranth (Grain and Leaf Versatility):**
   - Dual-use plant: edible leaves and high- protein seeds.
   - Extremely drought-tolerant and self-seeding.

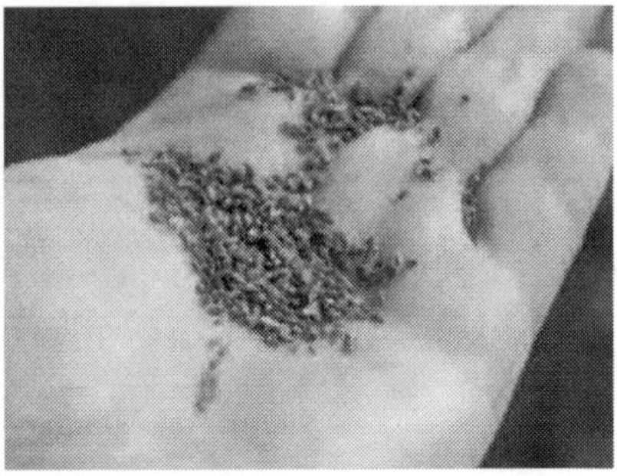

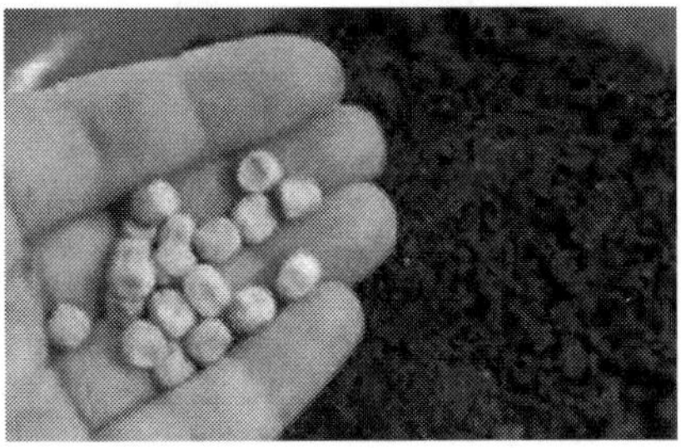

How to Stockpile These Seeds:

- Prioritize heirloom, open-pollinated varieties.
- Store seeds in moisture-proof, airtight containers.
- Keep in a cool, dark place.
- Rotate seed stocks every few years by growing and saving fresh seeds.

Planting and Harvesting Notes:

- Interplant species for mutual benefit (e.g., "Three Sisters" method: corn, beans, squash).
- Stagger plantings for prolonged harvests.
- Allow some plants to fully mature and go to seed for next season's planting.

Safety Measures:

- Keep seed stock diversified to protect against crop failures.
- Save seeds only from the healthiest, strongest plants.
- Avoid hybrid seeds for replanting unless no other option is available.

Integration Tip: Plant your critical seven crops near your greenhouse, solar dehydrator, and rainwater catchment areas. Use composted kitchen and garden waste to continually enrich your soil for these staple crops.

Benefits:

- Ensures survival nutrition in compact growing spaces
- Establishes a regenerative food system with seed- saving
- Reduces dependence on commercial food and seed supply chains
- Increases resilience against climate, economic, or political disruptions

Sustainability Focus: Focus on crops adapted to your local climate. Use organic growing methods and natural pest controls. Share seeds and techniques with neighbors to strengthen community food networks.

Chapter 54:
Comprehensive Tools & Materials Lists

Successful off-grid living relies on having the right tools and materials available when you need them. A well- planned inventory helps you build projects faster, maintain systems longer, and stay prepared for any challenge.

This appendix gathers all essential tools, materials, and supplies referenced throughout the book, organized for easy planning, sourcing, and checklist creation.

Basic Tools:

- Shovels, spades, and garden forks
- Hammers (claw, mallet)
- Screwdrivers (flathead, Phillips)
- Handsaws, hacksaws, and pruning saws
- Hand drills and bit sets
- Pliers, wrenches, and socket sets
- Tape measures, levels, squares
- Crowbars and digging bars

- Utility knives and scissors
- Wire cutters and crimpers
- Staple guns and heavy-duty staplers
- Buckets, basins, and tubs
- Funnels and strainers
- Trowels and seedling tools
- Hoes and rakes
- Paintbrushes and rollers (for waterproofing, sealing)

Advanced/Optional Tools:

- Power drills (manual or battery-powered with solar charging)
- Chainsaw or hand saw for timber work
- Solar panel installation tools (multimeter, crimpers, connectors)
- Welder (for metal frame repairs)
- Water testing kits (for filtration and catchment)

Common Construction Materials:

- PVC pipes and fittings
- Lumber (treated and untreated)
- Plywood sheets
- Roofing materials (metal sheets, shingles, waterproof membranes)

- Cinder blocks, bricks, and stones
- Concrete mix or mortar
- Rebar or metal stakes
- Wire mesh, fencing materials
- Hardware (nails, screws, bolts, washers, hinges, latches)
- Sealants, caulks, weather stripping
- Insulation materials (foam boards, natural fiber options)
- Tarps, rope, twine, and bungee cords

Food Production and Preservation Supplies:

- Seeds (heirloom and open-pollinated)
- Dehydrators (solar and electric)
- Canning jars, lids, and pressure canner
- Mylar bags, oxygen absorbers
- Food-safe buckets and barrels
- Compost bins or tumblers

Water Collection and Filtration Supplies:

- Rain barrels or large storage tanks
- Guttering systems and downspouts
- Sand, charcoal, and gravel (for DIY filters)
- Buckets, hoses, and drip irrigation components
- Water pumps (manual, solar, or low-voltage)

- Water storage containers (glass, food-grade plastic)

Energy Generation and Storage:

- Solar panels, inverters, and charge controllers
- Batteries (deep-cycle, lead-acid, or lithium)
- Wind turbine kits
- Bicycle generator components
- Fuel stabilizers and backup fuel storage tanks

Survival and Security Supplies:

- CB, FRS, and GMRS radios
- Flashlights, headlamps, and solar lanterns
- First-aid kits and natural remedy supplies
- Razor wire, tripwire alarm supplies, motion lights
- Fire extinguishers (Class A, B, and C)
- Personal protectivc equipment (gloves, goggles, dust masks)

Sustainability and Integration Focus: Whenever possible, prioritize salvaged, recycled, or repurposed materials. Invest in high-quality, durable hand tools that can be maintained and repaired over many years, reducing

dependence on disposable products and grid-based resupply chains.

Chapter 55: Project Variations, Modifications, and Additional Tips

Off-grid living isn't a one-size-fits-all journey. Terrain, climate, available materials, and personal needs vary widely from one homestead to another. The true spirit of resilience lies in building systems and adapting and customizing them to fit your unique environment and goals.

This section provides project variations, modifications, and practical tips to help you tailor everything you've learned to your specific situation, improving efficiency, durability, and integration across your property.

Water Systems Variations:

- **Rainwater Catchment:**
  - Use inverted tarps or wide funnels if gutter systems are impractical.
  - Install first-flush diverters to keep initial dirty runoff out of tanks.
- **Filtration:**
  - Upgrade to ceramic or UV light filters if running high-volume household systems.

  - Add sand or charcoal pre-filters to reduce debris and extend filter life.

Energy Generation Modifications:

- **Solar Systems:**

- Build adjustable solar panel mounts for seasonal sun angle optimization.
    - Use salvaged window glass to build homemade passive solar heaters.
- **Wind Turbines:**
    - Construct micro-wind turbines from old bike wheels and PVC blades for low-budget setups.

Home and Structure Tips:

- **Storage Sheds:**
    - Incorporate green roofs with shallow-rooted plants for natural insulation.
- **Greenhouses:**
    - Add thermal mass (water barrels, stone beds) inside greenhouses to regulate nighttime temperatures.

Defense and Security Options:

- **Motion Lighting:**
    - Install dual systems: motion-activated floodlights backed up by simple solar garden lights to maintain visibility

during grid failures.

- **Perimeter Alarms:**
    - Bury tripwire alarms slightly underground with a lightweight soil covering for stealthier setups.

Food Production Variations:

- **Aquaponics:**
    - Use old bathtubs, repurposed totes, or barrels if larger tanks are unavailable.
- **Gardening:**
    - Build keyhole gardens or hugelkultur beds for improved water retention in dry climates.

Preservation and Storage Tips:

- **Root Cellars:**
    - Bury insulated coolers underground for micro-cellar storage if full excavation isn't feasible.
- **Dehydrators:**
    - Incorporate black-painted metal panels to boost solar dehydrator efficiency in cooler areas.

Communication Adaptations:

- **Radio Systems:**
  - Use mesh bags filled with small stones to anchor CB antennas without digging postholes.
- **Blackout Preparation:**
  - Create "go kits" for emergency communication, including pre-programmed radios, maps, and check-in protocols.

Overall Tips for Adaptation:

- Always salvage before buying new materials.
- Build modular designs that allow easy expansion or repair.
- Blend projects together whenever possible to create synergistic systems (e.g., collecting rainwater off the greenhouse roof for aquaponics refills).
- Document your system layouts, materials used, and maintenance needs in a simple binder for quick reference and knowledge transfer.

Sustainability Focus: Prioritize regenerative

practices: compost all organic waste, recycle graywater, rotate crops, and continually seed-save. Every adaptation should aim to make your system stronger, more redundant, and increasingly independent over time.

Chapter 56:
Safety Equipment, Procedures, and Best Practices

While building a self-sufficient off-grid lifestyle empowers independence, it also requires a higher level of personal responsibility. Without immediate access to emergency services, staying safe through preparation, proper equipment, and best practices becomes absolutely critical.

This section outlines essential safety gear, daily protocols, and smart habits that protect your property, projects, and family — ensuring that your off-grid systems remain sustainable, resilient, and hazard-free.

Essential Safety Equipment:

- **Protective Gear:**
  - Work gloves (cut-resistant, chemical- resistant)
  - Safety goggles and face shields
  - Hard hats (especially for construction projects)
  - Steel-toe boots
  - Hearing protection (for chainsaws, generators)
  - Respirators or dust masks (for sanding, painting, filtering ash)
- **Fire Safety Tools:**
  - Fire extinguishers (Class A, B, and C)
  - Fire blankets
  - Buckets of sand or water at major workstations
  - Smoke detectors (battery or solar-powered)
- **Medical Supplies:**
  - Comprehensive first-aid kits
  - Herbal remedy kits
  - Trauma kits (tourniquets, clotting agents)
  - Emergency medical manuals or guides

- **Emergency and Communication Gear:**
    - Two-way radios, CB radios, or ham radios
    - Signal flares, whistles, and mirrors
    - Flashlights, headlamps, and backup solar lanterns

Best Practices for Safe Off-Grid Living:

- **Worksite Safety:**
    - Clear debris, tools, and obstructions daily.
    - Mark hazards visibly (holes, trenches, wires).
    - Always have a buddy or check-in system when working with heavy equipment.
- **Fire Safety:**
    - Keep brush and flammable debris cleared around buildings.
    - Create defensible zones around critical structures.
    - Store fuels securely away from living quarters.
    - Practice controlled burns only with proper clearance and weather conditions.
- **Tool Safety:**
    - Inspect tools before each use.

- Maintain blades, motors, and cutting edges regularly.
- Use correct protective gear based on the task.

- **Water and Food Safety:**
  - Test collected rainwater before use.
  - Sanitize food preservation areas.
  - Rotate stored food and water supplies systematically.
- **Medical Preparedness:**
  - Train in basic first aid and CPR.
  - Establish evacuation and emergency response plans.
  - Maintain updated emergency contact information in waterproof formats.

Integration Tip: Build "safety hubs" around your property: small storage stations containing fire extinguishers, first-aid kits, flashlights, and radios at key points like workshops, greenhouses, and living areas. This ensures rapid access to life-saving equipment when needed.

Benefits:

- Reduces accidents and injuries
- Increases system longevity through safer operation
- Enhances emergency readiness and resilience
- Provides peace of mind and confidence in daily work

Sustainability Focus: Choose reusable, durable safety gear over disposable versions. Salvage materials for firebreak construction or tool repairs. Integrate safety checks into your regular homestead routines, making preparedness second nature rather than an afterthought.

Conclusion

In uncertain times, it is important to note that self-reliance isn't just a dream — it's a necessity.

Everything You Need On Your Property gives you the practical skills, step-by-step instructions, and sustainable wisdom to transform any piece of land into a self-sufficient powerhouse.

Discover how to:

- Harvest and purify your own water
- Grow, store, and preserve your own food
- Generate off-grid energy from sun, wind, and water
- Secure your home and defend your property
- Craft natural remedies and emergency supplies
- Build essential structures using local, sustainable materials

Whether you're a beginner looking to start small or a seasoned homesteader expanding your systems, this book empowers you to design an independent life — resilient, regenerative, and ready for anything.

Your future doesn't come from a store. It grows, it builds, it lives — right here.

Reference sources

1. Falk, Ben. *The Resilient Homestead: Innovative Permaculture and Whole Systems Design for the Farm and Homestead*. Chelsea Green Publishing, 2013.
2. Hemenway, Toby. *Gaia's Garden: A Guide to Home-Scale Permaculture*. Chelsea Green Publishing, 2009.
3. Ashworth, Suzanne. *Seed to Seed: Seed Saving and Growing Techniques for Vegetable Gardeners*. Seed Savers Exchange, 2002.
4. Seymour, John. *The Self-Sufficient Life and How to Live It: The Complete Back-to-Basics Guide*. DK Publishing, 2003.
5. Jenkins, Joseph. *The Humanure Handbook: A Guide to Composting Human Manure*. Joseph Jenkins, Inc., 2005.
6. Strauss, Mark. *Survival Gardening: 25 Proven Methods to Grow Enough Food to Survive and Thrive*. Self-Reliance Publications, 2019.
7. Campbell, Stu. *Let It Rot!: The Gardener's Guide to Composting*. Storey Publishing, 1998.

8. Haeg, Fritz. *Edible Estates: Attack on the Front Lawn*. Metropolis Books, 2010.
9. Holmgren, David. *Permaculture: Principles and Pathways Beyond Sustainability*. Holmgren Design Services, 2002.
10. Kane, Jason. "Building DIY Water Filtration Systems for Off-Grid Living." *OffGridWorld.com*, 2020.
11. Permies.com. "Homesteading and Permaculture Community Resources." Accessed 2024.
12. Seedsavers.org. "Seed Saving Techniques and Heritage Seed Preservation." Accessed 2024.
13. The Grow Network. "Home Remedies, Emergency Food, and Backyard Food Production." Accessed 2024.
14. McElroy, Lisa. *Off the Grid: Modern Homes and Alternative Energy*. Gibbs Smith, 2012.
15. Mickelson, M.D., Reed. *Practical Survival: Preparing for the Unexpected*. Self-Published, 2017.

Made in the USA
Monee, IL
05 November 2026

67811019R00212